LANDSCAPE PAINTING
AND
THE AGRICULTURAL REVOLUTION

LANDSCAPE PAINTING AND THE AGRICULTURAL REVOLUTION

G.E. FUSSELL

The Pindar Press
London 1984

Published by The Pindar Press
66 Lyncroft Gardens
London NW6 1JY

British Library Cataloguing in Publication Data

Fussell, G. E.
Landscape painting and the agricultural revolution
1. Landscape painting, English
2. Landscape painting–18th century–England
3. Landscape painting–19th century–England
I. Title
758'.1'0942 ND1354.4

ISBN 0-907132-17-0

Printed in Great Britain by
Photobooks (Bristol) Ltd

Contents

Preface

Art is a small word which covers a multitude of gifts, literary, pictorial, musical, each bestowed upon the chosen. The pictorial art, the painted picture, is one of the most pleasing, the most immediate in its impact upon the senses. Much that is modern is entrancing because of its challenge and compulsion to work out its meaning. The pictorial art of an earlier time was representational, mainly beguiled into portraying scenes from life as the artist saw them. The pictures are in some sense photographic records of then current social occasions. The artists painted their contemporaries and their environment, their personal appearance, the clothes they wore, the furniture of their homes, the surrounding grounds whether landscaped or not, their horses and carriages, their cattle, sheep and pigs, the peaceful rural scene, cottagers and their cottages, their children, and so on. These things are in the most accurate sense social history, and bring to life the descriptions of the written word.

A modest caveat must be entered here. William Marshall, who lived in the great age of landscape painting, posed a delicate question and one that is unanswerable. "Did ever a landscape painter", he asked, "find a scene purely natural, which might not have been improved by the hand of art, or which he did not actually improve by a stroke of his pencil?" [1] Whatever may be thought about this observation, the great landscape painters of the eighteenth century had so far educated the perceptions of their contemporaries that the anonymous author of *An historical and descriptive Guide to Scarborough and its environs* was moved to remark that in that place "in an unengaged hour the judicious in landscape will find many a sketch of detailed scenery worthy of Gainsborough's charming art or Sandys' light and flowing pencil".

As a record of the modern appearance of rural landscape, photography has replaced the landscape painter. The modern world is full of pictures. The cinema and television tell their mundane stories visually, though there is dialogue. In an earlier day the passing scene, the portrait of an eminence, animal or human, was sedulously painted, as was the rural landscape and the rustic occupations of the majority of people and the world they lived in. Such pictures supply a visual memento of past times, art which has become history. The great age of this style of painting was equally a time when men's activities made a vast alteration in the appearance of the countryside, and some part of the record of this change is to be found in the work of the famous landscape painters, and animal painters, of the eighteenth and first half of the nineteenth centuries. The ubiquitous camera has today outmoded the painter because everybody, his wife, his son and his daughter can now manufacture pictures for themselves.

This does not invalidate the work of the great painters of landscapes and country scenes whose time has passed. Their work depicts the scene of their day, and it can be recaptured visually in no other way than by looking at their paintings. They supply a backward glance that the most vivid word picture must lack, correlative as these two methods of expression are. The paintings these artists made are no longer contemporary, but are history.

What are historical facts? This is an unanswerable rhetorical question. It will continue to be unanswerable because each pair of eyes sees things differently; but the picture of the event painted from nature is a permanent record of what was.

NOTES

1. William Marshall, *Rural ornament . . .*, London, 1795, I, p. 249.

I

The Birth of the Agricultural Revolution, 1660–1730. The Example of Low Countries Farming and Landscape Painting

New times often produce new ideas, a truism that modern events have heavily underlined: otherwise there would be no need, if indeed there is such a need, for new books presenting new interpretations, or making connexions between events which, at first glance, seem absolutely disparate. Hopefully what is presented here is the idea of a connexion between two historical events, contemporary with one another, that would at first sight seem only tenuously connected; whereas they had in fact a most intimate relation. It was not until the so-called agricultural revolution was fairly advanced that this relationship was defined by the run of contemporary happenings, but every human process develops from microscopic beginnings until it becomes nation wide, or even in our modern times, with the contemporary ease of communications, world wide.

Even in the times of our ancestors of three hundred years ago there was a great deal of international communication which overran the boundaries of states and even the brut of warfare. One of the most important of these from the English point of view was the result of the observation of Flemish husbandry by a visitor who spent some time in this area in the early seventeenth century.

Sir Richard Weston, of Sutton Hall, near Guildford, went to the Low Countries during the 1640s, for reasons that have yet to be decided. He brought back with him many farming suggestions that were completely novel to his neighbours and more distant contemporaries. He proposed to follow on his own estate the rotation of crops he had seen in Flanders. His example and propaganda was the slender foundation upon which English farmers of the next century or two built the new system of mixed farming that put them into the forefront of the farming world by the mid-nineteenth century. [1]

The influence of Sir Richard Weston has been emphasised because he wrote a book about the procedure as well as demonstrating it in practice on his Surrey estates: but at about the same time the money-making and fertility creating turnip crop was already being grown in isolated places in the eastern counties, for example at Weybread in Suffolk.

But no artist was moved at that time to paint a picture of a field of clover nor one of a field of turnips, both then very unusual features of the English countryside, which gave it a novel appearance in those places where either of these crops was cultivated. Enthusiasm for the appearance of the landscape was limited to the horrific, or magnificent, mountains, deserts and rocks. Improved farming was a skill that had not yet attracted more than a cursory attention, and, when the rural scene was immanent in almost everybody's life, it was neglected in favour of prestige art, such as portraits that could be hung in sequence in the great halls of the nobility, and thus disclose the family resemblances in the pictures of the successive generations, a sort of pictorial studbook. The rural scene, interesting and changing as it slowly was, had not yet aroused interest among the wealthy who could afford to buy pictures, or have them painted, in England. Family portraits and family prestige was the thing; but these were most usually portraits of the most elevated among the nobles and aristocrats. Neither the great landlords, nor the lesser, or only a very few of them, were at this time engaged in transforming a part of the face of England, and so making it a suitable subject for pictorial art both before and after its transformation. The new scene created was a fair and pleasing vista of itself in its reality, and pleasingly suited to pictorial art.

Progress, or rather change in farming systems, cannot properly be measured by the appearance of new textbooks on the subject; but nobody would write books unless they at least hoped for publication and in consequence readers. These readers must be interested in the subject treated: it must in fact be one that is in the forefront of contemporary life. In the late seventeenth century subjects of common interest were neither so numerous nor so varied as they are in modern times, nor could the focus of interest be diverted or distracted because there were no news media to fragment thinking into minute particles.

Since our ancestors' outlook was so much more restricted, and confined to the immediate problems of living, the most important of which was food production, they may have been more ready to read textbooks of farming, the occupation by which the great majority of them lived, directly or indirectly, landlords, their tenants and the tenants' employees. So it may be assumed that the textbooks were of a kind that might (though some of them were not) be acceptable to the probable readers, and by being acceptable would reflect the activities, the daily round, and any modifications that could be recognised as likely to increase production, and thus add to the producer's income. Here therefore the writers of contemporary textbooks will be used as references to what was happening. Possibly not so detailed as the minute investigation of documents relating to individual farms, conveyances, leases, wills and inventories, these textbooks can be regarded as a reflection of contemporary farming, and as advice, whether good or bad, how to make more money by working in a different way on a different system.

There were not many such works in the late Stuart age. The most important were written by Worlidge, Yarranton, Tull and Mortimer: though Tull did not publish his works before 1731, his experiments, mechanical skill and theories were discussed freely in the period. All these writers would have been classified in the social system of the day as gentlemen, a description they would have proudly acknowledged.

It will not be at all novel to discuss them and their works, but it seems to be necessary, if only in brief. John Worlidge described himself on the title-page of his *Systema Agriculturae; the mystery of husbandry discovered . . .*, as J.W., Gent., and stated that the book was published for the common good. The first edition, a folio, was issued in 1669, and successive editions appeared with some frequency during the rest of the seventeenth century. It was embellished with a frontispiece showing most of the work required during the year, was provided with a rustic calendar of operations and a dictionary of "Rustick Terms".

No complete details of Worlidge's life can now be discovered, but it is known that he lived at Petersfield, Hants., and there can be no doubt that he must be classified socially as what used to be called a gentleman, that is a landed proprietor, a country squire; but he was probably one of those whose life followed the expected round, and was so unremarkable personally that no curiosities of character, oddities of behaviour, or accidents having an anecdotal smack can be recorded of him. He was an assiduous writer, and the evidence of his productions makes it clear that he was widely read in the literature of his subject, both classical and contemporary. But unlike most of those whose activities in the study are more remarkable than their activities in the physical world, Worlidge managed to combine his literary work, which was extensive, with some attempts to figure in the practical world of the farm fields. He was a strong advocate of the so-called new crops, lucern, sainfoin, clover, rye or rye grass, and turnips, and, being of a mechanical as well as a literary and agricultural turn of mind, he designed a seed drill, something that had been discussed half a century earlier by a man named Gabriel Plattes. He produced a pictorial design of this machine, and a model based on his theories has been made in recent years, but the machine did not come into production at the time, and the idea remained latent for some years. Worlidge also wrote about horticulture and cider, but these subjects may perhaps be neglected for the moment. [2]

The outstanding novelty of Worlidge's writings is perhaps his attempt to design a seed drill, in which he was a forerunner of Jethro Tull. His recommendation of the 'new' crops was not a novel and personal achievement. It had been made since the sixteenth century, by Googe amongst others; but it is doubtful how soon this recommendation was acted upon, except by an isolated enterprising individual. Walter Blith had made the same recommendation during the Commonwealth, but the most formidable protagonist of clover was Andrew Yarranton of Worcestershire, who wrote a few years before Worlidge. He brought clover seed to Ross in Herefordshire, and popularised its cultivation in Worcestershire and the adjoining counties between 1653 and 1677. Landowners were impressed, and some cultivated the crop with profit. For example, Sir John Winsford, a former High Sherrif, put no less than 60 acres under the crop as a result of Yarranton's propaganda. There were others, some of whom did not make a success of the crop. It was not only in this predominantly grazing area, the West Midland counties, that clover was grown. Nicholas Toke, a landowner in Kent, bought clover seed in 1653, and from that year his accounts show that he was among the small number of adventurous, enterprising landowner farmers who grew clover as a commonplace crop. He also cultivated turnips as a fodder crop. Turnips were also being grown on the borders of Norfolk and Suffolk at Weybread, as Eric Kerridge found. [3] Most of this progress, sporadic as it was in the late seventeenth century, has been indicated in an essay of my own.[4] Kerridge has described the making of water meadows in the southwest during this century, installations that were designed to provide a larger supply of forage.

But these new crops and ideas, important as they were, and they were to make a major impact upon agriculture and animal breeding in the eighteenth century, were subsidiary to the adventurous spirit and inventive mind of Jethro Tull, another gentleman, widely travelled and educated in the most superlative style then possible. He was the designer of the first practical seed drill, and was by so much the iniatiator of mechanised agriculture; but his life and work have been described in a recent work of my own, and need not be gone into in any detail here.[5]

The people responsible for these innovations in the field are hard to find, but possibly exploration of county archives will disclose more of these important pioneers, who, at the remote distance of nearly three hundred years, have exercised a decisive influence upon the population and their lives in modern times. Jethro Tull was one of these, whose direct influence our mechanically minded world is only too ready to accept. Other landed gentry made their trials and experiments in the field, but not often in the field of mechanics, with less enduring a reputation. Few others in the period under discussion have emerged from obscurity. One of the best known was Edward Lisle, a landowner of properties scattered over several counties, an educated man, in the category of what is now called forward looking.

Lisle lived at Crux Easton, Hampshire, near where the three counties of Berkshire, Hampshire and Wiltshire abut upon one another, but his properties were dispersed over the country from Northampton to the Isle of Wight. As he went to and fro between these places in the later years of the seventeenth century Lisle made notes of the farming practices of his tenants and others. These notes were organised and published as *Observations on Husbandry* by a descendant in 1757. Since the information describes what named farmers were doing in specified places, Lisle has provided excellent material for the historian, as valuable to the modern reader as it was to the contemporary seeking information about the best practices of his fellows. Lisle was not one of the numerous crowd of husbandry writers who offered panaceas of startling originality that would make the fortune of anyone who assimilated and adopted them. His words were the words of wisdom of contemporary practical men. [6]

This practical wisdom did not exclude the cultivation of the new, or so-called new, fodder crops, the grasses, legumes and turnips. Migrants came from Wales to hoe the last crop. It is now known that many seventeenth century farmers, who lived along the chalk ridge of the Chilterns and Berkshire Downs were growing sainfoin even in the open fields of that area. Clover and rye grass were crops in widely separated districts in addition to the West Midlands where Yarranton had promoted these feeds. Another way of supplementing fodder supplies was the construction of water meadows, more especially in Gloucestershire and Wiltshire. These added, as Kerridge has pointed out, to the resources for feeding the farm livestock, an addition which was the more welcome in days when natural grass and wild herbage was almost all that was available. [7]

John Mortimer, like so many others who became landowners, was a London merchant; but when in 1693 he became owner of an estate, Toppings Hall in Essex, he devoted the rest of his life to farming it. Mortimer was one of the long succession of famous Essex farmers, but his name is not as well known as, for example, that of Thomas Tusser or Alderman Mechi. Mortimer had other outstanding achievements. He married no less than three times, a daring undertaking, but one worthy of a descendant of the Somerset Mortimers, whose estates are said to have been washed away by the sea, a less preventible catastrophe than having them washed away by unwise and uneconomic agricultural experiments, valuable as they may be to posterity. [8]

Mortimer's book, *The Whole Art of Husbandry*, published in 1707 in two volumes, has some value, apart from whatever use it may have been to his contemporaries. It describes, sometimes with approbation, what farmers and landowners were doing in many different places scattered about the countryside. Like his Essex predecessor, Thomas Tusser, he was all for enclosing land "which lays a foundation for industry and good husbandry, because of the security it gives a man in the quiet possessing the benefit of his labour and care". This was very true, and had long been realised by advanced thinkers: enclosure was to become a major preoccupation in the second half of the eighteenth century. Mortimer was a writer intrigued with novelty as were so many other writers then as later, but the book is, or rather has been assessed as, marking "an epoch in agricultural literature . . . The arrangement (of the subject matter) is more systematic than some of the earlier works". [9] Needless to say Mortimer was a strong protagonist of the currently popular "improvements" that were to become more generally acceptable as the eighteenth century advanced. It is perhaps unnecessary to set them out in detail here.

It had long been the fashion amongst the members of aristocratic families, like those of Tull and Lisle, for the current representative to have a portrait painted. The pictures of the living members of the family, more especially perhaps that of its head, were usually hung in place in one of the galleries, and the place was in order of succession to the dead ancestors who graced the place of exhibition. Both Tull and Lisle conformed to this procedure, and had their portraits painted, but there is no evidence to hand to indicate that their aesthetic appreciation and taste led them to purchase or commission other pictures. Nevertheless the existence of these status symbols confirms the social position of these two men. Neither Andrew Yarranton nor John Mortimer seem to have competed with the others in the matter of having their portraits painted, though they were members of the same social class.

It would be a great deal less than desirable to go very much further in a discussion of the farming textbooks written at this time. Such a procedure would be only to repeat what was said in my *Old English Farming Books, 1523–1730*, as long ago as 1947. My current remarks will be restricted to one other writer, Richard Bradley, about whom controversy has raged from his times to our own. He certainly made some contribution to knowledge in his researches into what are now recognised as virus diseases of plants, and with his equally significant researches into plant physiology; but, combined with his undoubted abilities in these directions, there were other activities by no means so commendable though this is not the place to delve into them.[10] He must, it seems likely, have made a not insignificant contribution to the subject, not only in his own day, but for posterity.

These and many others less articulate were the background of such agricultural progress as there was in the late Stuart period. It was considerable if measured in the context of the time, something that has been emphasised by recent writers. These men were not the Grub Street hacks that their predecessors and successors, the writers of agricultural textbooks, are often scathingly called by their contemporaries and later pundits. They were articulate members of a class already beginning to be concerned with the productivity of their estates. They certainly did belong to the class that was educated along the then conventional lines. They were not to be classified in the social scale as nobility, but they were most certainly gentry, or what may be called the subordinate or subaltern branch of the ruling class. After all few others were educated in the classics, though some, the helots of the system, must have been able to figure because of the growing need for people who could keep accounts.

To some extent agricultural theory at this date was international: it becomes more so as the years of the eighteenth century rolled by. The predominant influence was that of the Flamands and Netherlanders. They were the example for Europe, and this was not limited to their skill in farming processes that were probably the most productive in Europe at that time, and for a century or more thereafter.

In the Netherlands business, too, flourished, especially foreign trade, and with growing riches the natives wished to adorn their homes in any and every way. They were immensely successful as gardeners, making floriculture a fine art, and this may have been ancillary to the development of the Dutch school of painting that later exercised so great an influence in our own country. There were earlier influences. In every human activity, especially the literary and pictorial arts, there is a continuous tradition leading back into the remote past. The pictorial arts lend themselves readily to comparison if only because they are pictorial. The line of descent from remote times is less immediately apparent in literature, where it must be traced by mental effort, and is even more tenuous in relation to what are known as artefacts.

The remotest past is shrouded in shadows, and here all that can be done is to suggest a clue to the continuing interest in their own activities taken by the human race, so that individuals feel or felt urged to put those activities on record pictorially, or to describe them in the written word. The major pre-occupation of mankind until quite recently was food production. War and the saving of souls were minimal in the employment they provided, only a small proportion of men being engaged in these profitable branches of work. Merchanting and manufacture were growing from the Middle Ages, and were occupying more and more people as time passed and the modern world slowly came to birth.

Since the farmers of the Low Countries, and Flanders in particular, practised the most developed and productive systems in Europe in the sixteenth and seventeenth centuries, no-one need be astonished to find that the native painters in these countries occupied themselves in depicting scenes of rural life. Pieter Bruegel (1525–1569) was one of the earliest of them; and his love of country scenes and work is exhibited nowhere better than in a series of the Months. Unfortunately only some of the set remain (Pls. 1–2). No-one can say with certainty that these are in some sense derivative, but in the Middle Ages it had been customary for illuminators of the calendars often prefacing Psalters and Books of Hours to define the months by pictures of the characteristic occupation of each one. Many of these are very beautiful, and some may have been part of Bruegel's aesthetic inheritance (Pl. 3). Possibly one of Bruegel's best known pictures is *The Return of the Herd*, showing cattle being driven down from the hills where they had summered to winter quarters in the lower land (Pl. 4).

These pictures have no religious overtones. They are purely rustic scenes. Many of Bruegel's successors painted the same sort of pictures, not, it may be added, the occupations of the months as such, but scenes of rural life, the pleasures of the people at games of bowls, skating the orgy of the kurmess, and so on, created for a secular and aristocratic patronage rather than an ecclesiastic. Most of these artists, like Bruegel, had visited Italy, and those who had not done so were not immune from the influences emanating from that country. The consequence was the continued attention of artists to scenes from the Bible, or from classical mythology. "Susannah and the Elders" seems to have been a favourite subject. These incidents were not painted *in vacuo*, but the figures, statuary and buildings were enclosed in landscape. Roughly speaking this was the landscape of the time. St. Sebastian, pierced with arrows, is often shown in his death throes against a background of smiling countryside. This is not surprising because almost everybody lived in such surroundings, even in Italy.

The landscape paintings of seventeenth century Italian artists were inclined to the horrific, based upon the current feeling that mountains were terrifying. Salvator Rosa painted monstrous scenery in ghastly and forbidding colours (Pl. 5). Poussin was no less informed with this spirit. Titian and Tintoretto were occupied with the same style, which has been called "a sense of grandeur and noble scenery", but the most important was Claude, who is said by an admirer to have produced a complete picture of nature as a whole, while the others, with all their magnificence, "had a certain remoteness from pure nature", a judgment with which it is impossible to disagree. These men certainly had a marked influence on aesthetic taste in the eighteenth century; but the Dutch and Flemish painters, influenced as they doubtless were by the great masters who went before them, looked at life in a more domestic way, in the daily life of the people amongst whom they lived. Though they had also been subjected to Italian influence this did not have a marked effect upon the character of the productions of the Dutch school, who employed themselves in depicting their local rural occupations and scenery with some emphasis on domestic interiors.

These artists were very numerous, and it would be impossible, even if desirable, to catalogue them here to support the contention that the seventeenth century in these countries was a precursor and exemplar to later English landscape painting, the latter being inspired not only by the poetry of the rural scene itself, but by the changing appearance of a great deal of their native countryside.

A few names must be mentioned, if only as examples. Rubens (1577–1640) produced some vivid landscapes. Frans Hals (?1580–1666), Rembrandt (1606–1669), van Dyck (1599–1641) are amongst them; but perhaps the most important from an agricultural point of view were Paul Potter and Aelbert Cuyp, whose rural scenes and careful portraits (no lesser word is appropriate) of animals in their domestic setting, the field and barn, are very likely to have influenced English landscape painting of the eighteenth century though patriotic commentators would like to believe that the latter was an indigenous product developed by the economic and technical circumstances of the century.

The Dutch and Flemish schools continued to be active throughout the seventeenth and in the beginning of the eighteenth century when they began to fade out. David Teniers (1610–1690) was a good example. He painted the people at their work and recreation in surroundings of fine landscape. Pynacher (1622–73) made scenes of country life, shepherds and so on.

The nexus between these painters and their native farming community is very simple. It is only that the pictures depicted the farmer's activities at work and play. The farmers worked in the physical world of the scenes the artists reproduced in their pictures, possibly for the benefit of the townsmen, merchants and others, who had made fortunes in foreign and domestic trade, but who were not far removed from these country sights. A man ploughing in a field seen by a passer-by, and a picture of a man ploughing in a field look much the same in the eyes of the observer if the picture is well designed though the living scene is animate and the pictorial one inanimate. The former is evanescent as the passer-by moves on; the latter is permanently before the eyes of its owner. Who bought the pictures and why is not immediately relevant here. What is to the point is that the pictures existed, and, since Rome was visited by many of the later English landscape artists, there can be very little doubt that they made themselves familiar not only with the Italians but also with the great painters of the Dutch and Flemish schools, and possibly the lesser lights, though this is not proven.

Some of the Low Countries painters came to England, Rubens for example, but he did not paint the English scene, although he was received at the court of Charles I. Indeed Inigo Jones (1573–1652), the famous artist and designer of masques, was a landscape painter, but the English scene did not find favour at that time. It was possibly too familiar, homely and domestic. This denigration of the homeland was before the agricultural revolution and its accompanying prosperity had become evident in the land. It was before the time of the literary farmers of the period 1690–1730, whose ideas have been described. In many of the foreign pictures there was a greater emphasis on the country scene itself rather than upon the people who figured in it. Often these people were small and insignificant in the design, and in some pictures were omitted altogether.

The topographical artists like Hollar (1607–1677) and Place (1647–1728) are regarded as representing the beginning of the appreciation of landscape in England. Their works were doubtless widely dispersed amongst the limited intelligentsia of the time. Another man who depicted the English countryside was Francis Barlow (c. 1626–1704), his picture of "Partridge stalking" being something that might be acquired by a rich landed sportsman. Incidentally the stalking was done with the help of a stalking horse. Place's picture, "The Dropping Well at Knaresborough" (Pl. 6) is rather horrific, but then the place itself must have been forbidding to the seventeenth century travellers and invalids who visited it with so much difficulty and danger. Thornhill, who was born at Melcombe Regis in 1675, and lived until 1734, made pictures of great houses naturally surrounded by landscape. Topsell in his *History of four footed beasts and serpents* . . ., 1658, used illustrations of the various animals he described and made good drawings of them.

But it has been said that the beginning of natural landscape painting in England was made by Richard Wilson, and with him the intimate relation of landscape in pictures and on the ground itself became important in England, as important indeed as the improved farming that made it possible for the gentry who owned most of the land, and certainly ruled it, to buy finished pictures by the remarkable artists of that age and time.

NOTES

1. On Weston see my *Old English Farming Books*, Vol. I, 1947, pp. 40–43, 54, 76, 115 and 128.
2. *Ibid*., pp. 56, 68–72, 95–96.
3. E. Kerridge, *The Agricultural Revolution*, George, Allen and Unwin, 1967, p. 270.
4. G. E. Fussell, "Pioneer farming in the late Stuart age", *Journal of the Royal Agricultural Society of England*, 100 (1940) pp. 13–19.
5. G. E. Fussell, *Jethro Tull, his influence on mechanised agriculture*, Osprey, 1973.
6. On Lisle see my *Old English Farming Books*, Vol. I, 1947, pp. 94–95.
7. Kerridge, cited note 3, pp. 251ff.
8. On Mortimer see my *Old English Farming Books*, Vol. I, 1947, pp. 96–98.
9. Mary S. Aslin, *Catalogue of the Rothamsted Library*, 1926, 2nd ed. 1940, s.v.
10. See my *Old English Farming Books*, Vol. I, 1947, 106–113.

II

Landscape change, landscape gardening and painting

During the thirty years from 1730 to 1760 there was a marked acceleration of change, a social phenomenon that was to become ever more rapid as time passed. There is supposed to have been a reason or reasons for this, but it is difficult if not impossible to formulate these changes in terms of human psychology, or in any other simple way. There were many impulses, animate and inanimate, but the most sceptical must admit that most of them were human, despite some conditions seeming to change without any recognisable impulse of any kind. The most important was the growing numbers of the population, for which many explanations have been offered, none of them more than partly convincing.

It is common knowledge that no Census of Population was taken before 1801, but Rickman made some theoretical estimates of the number of people in the country at different dates during the eighteenth century. His figures are the foundation of most of the estimates made by later historians. [1] One of these, chosen quite at random, but doubtless as reliable as any other, because it is quite clear that nobody can be certain, is that in 1710 there were in England and Wales about 5½ million inhabitants; in 1740 some 6 million; in 1770 about 7½ million. The populations of Scotland and Ireland were relatively much sparser. During the last thirty years of the eighteenth century the rate of increase was more rapid, a point that will be returned to.

The accepted reasons for this development have been stated many times. They are, it is supposed, for it cannot be indubitably proved, that there was a decline in the death rate, especially infant mortality. It has been suggested also that people were marrying at an earlier age than formerly, and thus more children were being produced, a very difficult proposal to substantiate. But the physical habits of the people had not changed. Personal hygiene was not more practised. The person was not too frequently subjected to soap, of which there was very little, and water. The heavy woollen clothing worn was infrequently washed, if at all, but was worn to destruction. The coming of the Lancashire cottons did something to change this insanitary way of life as time passed. They were easy to launder.

The plague had vanished, but there were other devastating diseases such as smallpox though this was slowly being overcome. It was possibly more dangerous for a sick person to enter a London hospital than to remain at home. However squalid the home might be, conditions in the hospitals were most likely to be much worse. Standards of purity in purchased foods were non-existent. Poisonous dilutents were not infrequently used. [2]

The inhabitants of London, usually referred to as the mob, were rude and unruly, living in crowded and filthy tenements. Even the rich, who were spreading north and west, did not escape the prevailing lack of sanitary measures. Reasonable personal safety after dark in an unlit town was only comparative, and confined to the main streets. In the minor ones it was non-existent, and going armed was essential for personal protection. These minor streets were narrow twisting lanes and alleys festering with the offal that householders cast out and nobody bothered to remove. Houndsditch was properly named after the dead dogs cast away there. This was one of the more respectable of Cockney habits. The butchers, who probably slaughtered the beasts they bought, threw out the waste products "guts and all nastiness" into the streets where this material rotted away and stank to high heaven. Householders were as careless with household waste of every kind. It was simply discharged into the gutter where there may have been some running water, but where it lay because nobody bothered. Our ancestors must have been dowered with an indifferent sense of smell! They were apparently immune to the animal droppings from cart horse and carriage horse and the crowds of cattle, sheep and pigs that were driven through the streets. This is not the place to discuss the night life of the Metropolis, but it was incredibly dangerous even by modern standards. A modest statement is that it was very unsafe to go out alone or unarmed after dark.

There were only one or two towns of any size besides London, such as Bristol and Norwich, and by modern standards these were very small. The population was spread about the countryside, possibly more densely in the East Midlands, East Anglia and the south, but in all this part of the kingdom there were tiny villages whose denizens farmed, or were employed in ancillary occupations, many of them part time. The blacksmith, the cobbler, the carpenter and the mason usually occupied some land which they cultivated for subsistence. Over all in consequence was the squire, who was in the main little different from his father and neighbours, but who enjoyed magisterial authority. His great value was his interest in rural pursuits, his livestock, his land, his tenants, his own locality. He was a stranger and a country bumpkin if he ever visited London, but he was an outstanding personality in his own neighbourhood. Outside this area the social system was the same, but the countryside was much less, or perhaps it would be better to say, more thinly populated.

The squires who owned the smaller estates were modest replicas of the great landlords, the leaders in the early stages of the period of innovation in farming systems, more especially on the lighter soils. "Turnip" Townshend is known to every student of history from the schoolboy to the scholar. He began to emphasise improvements that had already been commenced on his estates at Rainham before he retired from politics to the country where he and others began to dig for marl to improve the texture of their light lands, to grow turnips and clover, and to feed thriving flocks and herds. Other great landowners, their relations and neighbours were doing the same kind of thing. Walpole, if he had not been so successful in politics, might well have been nicknamed "Turnip" Walpole because the new crops were grown on what seems to have been a not inconsiderable scale on his estates, by his father, and by other relations and neighbours. Rich farms were made by these people out of lands that had produced little before being improved.

No difficulty was encountered in these places because the land was not occupied in open field; but landowners who wished to emulate these remunerative endeavours were handicapped in many places by the occupation and perhaps ownership, though the word is difficult to be precise about, of the land in the scattered strips of the open fields as it had been possibly laid out for a thousand years. These aristocrats were no more troubled by the rights of their humbler neighbours than their ancestors had been when acquiring control of a section of the landscape. They got Private Acts through Parliament giving them powers of enclosure and redistribution of the fields, and so changing the face of our countryside. This procedure was no more and no less righteous than any other decision that depended upon power, but it was quite effective in giving more power to the mighty, a not unusual result of human actions designed to further that end. Between 1700 and 1760 some 152 private enclosure acts were passed dealing with some 237,854 acres of open field, but including some waste land: 56 other acts authorised the enclosure of 74,518 acres of waste during the same sixty years. About 300,000 acres is a very minor proportion of the 36 million acres of England and Wales, but this very small beginning led to greater things. Probably some injustice was inflicted upon the humble at this time, but it was legal injustice, and caused greater outcry at a later date. "All evils (indeed)" as Defoe said, "are to be considered with the good that is in them". It is not a point to be discussed here, because another matter, the resulting appearance of the countryside, is the immediate consideration.

It is quite certain that the enclosers suffered no pangs of conscience, which are rarely felt by persons whose pockets are being better filled, and who possess an ineradicable feeling of superiority to the rest of mankind. They were better clothed, better fed, often to repletion, and immensely better housed in their insanitary mansions than the rest of their fellow countrymen, and, more important for the present purpose, were sufficiently wealthy to be able to buy pictures. "The social structure of pre-industrial England had only one really important division", wrote Professor D. C. Coleman, "between those who were Gentlemen and those who were Players. For all the elaboration of status and hierarchy distinguished in the seventeenth and eighteenth centuries the line of greatest consequence was that older division which distinguished gentility from the common people", and gentility was then conferred by the possession of a landed estate especially one that had been in the family for as many generations as may be, and certainly more than one for preference.[3]

The pleasure garden was always a popular perquisite of such people, from the remote times of the hanging gardens of Babylon, really far too remote to be of consequence here; but what was of consequence was that these gardens were usually enclosed for privacy, as a variety of illuminated manuscripts of medieval times depict. Some must have been of large area. Bacon wrote of a garden of thirty acres as if it were not especially grandiose,[4] but whether such a garden was enclosed or not does not appear. Most of the late Stuart gardens were enclosed by high walls, and it was not until the early eighteenth century that the fashion changed. One Bridgman invented the Ha Ha, [5] and when the walls were pulled down in favour of this novel boundary, landscape art was introduced into the empirical world. It was landscape art on the ground, not on the canvas.

One of the most renowned of these landscape gardens was that at Stowe. It was initiated by Bridgman, followed by William Kent and continued by "Capability" Lancelot Brown. [6] Lord Cobham continued the work until by 1755 the garden covered some 500 acres, an area that, had it been farmed, could have made its owner wealthy: but Lord Cobham was not short of funds. Like some other members of the nobility he was able to remove the people who lived at Stowe in order to enlarge his "garden", and to house them in the adjacent village of Dadford. This was by no means a singular event. In the 1760s Earl Harcourt removed the inhabitants of Nuneham Courtenay to cottages situated on the Oxford Henley road. The landscaping of the Milton Abbey estate in Dorset involved the destruction of cottage dwellings that would have offended majestic eyes, and the building of a remarkably nice village on another site, though unfortunately the cottages there very rapidly became overcrowded. The then Earl of Devonshire did the same thing at Chatsworth though one stalwart resisted him and his cottage still stands in its original position. Examples could be multiplied. The process is almost as modern as that indulged in when making arterial roads. Some of the avenues, prospects, etc., were carried out over miles of country, but such magnificence was rather caricatured by Pope at Twickenham, Horace Walpole at Strawberry Hill, and Shenstone at the Leasowes. It was landscape gardening *in excelsis*, carried out for the benefit of a limited class, who not only owned but ruled the nation.

One of the intriguing things about the practice was the use of clover seed to make the large areas of grassland. A couple of examples will suffice to show the value of clover was appreciated not only by the farmers, or at least a few of them, but also by these artist gardeners. Wenham Coke, predecessor of the famous Coke of Holkham, employed Brown, who used several hundredweight of white clover seed and Suffolk hayseed in 1762 to make the very extensive lawns round the house. At Milton Abbey 1,120 lbs. of Dutch clover seed was used between 1762 and 1770. Clover is not appreciated in suburban lawns today, but perhaps the lawns of these great houses were cut for hay. Brown, wrote Dorothy Stroud in 1950, [7] transformed thousands of acres of English countryside from bog and heath or rough pasture into the magnificent parks which were one of the eighteenth century's most valuable contributions to our heritage, and a striking metamorphosis of the landscape. The question might be posed, was no productive land used for this enterprise?

Stowe has fallen from its great estate to become a public school, and the great landowners who were responsible for all this work have long since passed away. The role of the landed gentry, the peerage, owners of all these things, has infinitely changed. Instead of governing the country with autocratic powers – in spite of struggles for position and place – the descendants of these men still hold lands and titles (or some do), but have sunk to the status of public entertainers, collecting money for showing their homes, and for amusements like traction engine rallies or open zoos of wild animals not indigenous to Great Britain. It does credit to their inherited acquisitive powers.

The creation of aesthetic scenery by the landscape gardeners was reflected to some extent in the character of contemporary verse, which was another recreation of the literati, the well-to-do who had been educated in the standard pattern of the time. There had always been a tradition of pastoral verse descending from the pattern of classical times, Theocritus, Virgil, and so on; the English productions were far from realistic. The poets lived in a cloud cuckoo land that was purely imaginative, as any reader of such works as those of Spenser must realise. The eminent poet of the day was James Thomson, who was accused by Leslie Stephen of "trying to write a half serious imitation of Spenser". [8] Pope before him had written "Windsor Forest", and an essay on pastoral poetry, but Thomson did practice the nascent realism that was to fructify in the work of later poets. The sequence of publication of his "Seasons" was not that of nature. "Winter" came out in 1726; "Summer" in 1727; "Spring" in 1728; and the complete year, including "Autumn", in 1730. Few can escape the shackles of tradition. Thomson was not among them, any more than the *Spectator* was in its fantastic sketches of the activities of Sir Roger de Coverly, an early eighteenth-century country gentleman of a kind that existed only in the minds of metropolitan wits. Thomson was a country boy who became a city dweller, which is not to say that he did not carry country life in his mind, or even in his physical view, for it abutted upon the then narrow confines of his London. A simple quotation should suffice to demonstrate the point. In *Spring*,[9]

> Joyous, the impatient husbandman perceives
> Relenting Nature and his lusty steers
> Drives from their stalls to where the well-used plough
> Lies in the furrow, loosened from the frost.
> There, unrefusing, to the harnessed yoke
> They lend their shoulder, and begin their toil

Leaving the plough exposed to the weather all the winter was extremely bad practice, but modern farmers persist in doing the same thing with much more expensive machines, so it is clear that Thomson was intimately acquainted with the world around him as well as with his poetic ancestry. The ploughman figures in other poems, e.g. Gray's *Elegy in a country churchyard* of 1750, though this is only indirectly pertinent to the present purpose. More directly to that end is Dyer's *The Fleece*, which Henry Morley called "an elevation of the Georgic to the praise of commerce".[10] Dyer had produced *Grongar Hill* in 1726, which is something of a landscape poem.

Shenstone, as already said, tried to lay out his limited estate in a *multum in parvo* kind of way. Instead of farming his small estate of £300 a year as a commercial undertaking he wasted its resources in an attempt to make it into an ornamental ground on the lines of the great landscape gardens being created elsewhere. He made gardens, grottoes, temples and inscriptions after the best and most grandiose examples (Pl. 7), but let his house and land go to ruin. He has been praised as the originator of the *ferme ornée*, but all this is now a golf course.[11]

People who spent money like water on their gardens were not likely to wish to inhabit any residence of less grandeur than its surroundings. A spate of building accompanied the laying out of the landscape gardens. As all physical things are liable to change, the great landowners, who had all this work done, also had it recorded in pictures by the topographical artists of the day. These were not yet landscape painters any more than the great masters of the earlier schools were, though landscape played a part, necessarily, in their productions.

The majority of the squirearchy were by no means so magnificent as the great landowners who built palaces like Castle Howard. Blenheim was built at the public expense, but that does not detract from its splendour. Longleat, Woburn Abbey, Holkham Hall, and many other such places could be named, but the smaller manor houses, often little more than large farmhouses, were occupied by the minor gentry, who were, not to refine, a rather coarse, uncouth lot. How could they be otherwise? There is a grotesque difference of opinion about the situation of the rural population, then the majority of the kingdom's inhabitants. "In the first half of the eighteenth century the poorest labourers enjoyed a margin above bare subsistence (in Nottinghamshire), and with the exception of the arable farmers working on a small farm between 1730 and 1750 their social betters enjoyed a high level of prosperity". Seasons were kind, yields were good by the standards of the day, and tenants and landlords alike were comfortable.

There was little to choose between the squire and his tenants. Their daily work was on their land, whether it was arable in open fields, or enclosures in separate occupation. Countless other jobs were required in the course of looking after livestock. No doubt the squire and the parson did not do *all* the manual work personally, but some certainly did, *videlicet* Parson Trulliber.[12] Unless they worked, there was little else for them to do. What leisure these people had was spent in hunting in a relatively unorganised way, shooting, fishing and so on, and as a result of all this physical activity, in eating vastly and drinking more than could be thought possible. The village ale house was the social centre for the men; card parties, working in the still room, making embroidery and the care of children was all that there was for the women, who lived in the most shocking isolation, or so one modern judgment has it. They were practically housebound in the winter, but as they were not aware of any different circumstances they most likely accepted their lot as everybody does in every age, with or without complaint. The village was their home, and they feared and hated any stranger. Such people had little or no aesthetic feeling. Their lives were mainly devoted to producing the food and equipment they required to go on living, the exceptions being the day of the annual or semi-annual fair, the village fete and church ales that broke up the monotony of their days.

These middling landowners, members of the minor squirearchy, though they performed a variety of useful functions in local government and so on, were not yet in the class that bought pictures, not even portraits of themselves. The greater landlords, who did buy pictures, looked upon the lesser fry with some sort of aimable contempt. They probably tried the same tactics on the painters whose pictures they bought. As any reader of the Dictionary of National Biography and other relevant works knows well enough, the artists were often men who had been born in the artisan class. It was no novelty for an artist to graduate from a boyhood training in coach painting and other kinds of decorative work into the somewhat superior grade of a painter of pictures of one sort and another. The grandees were obliged to buy the products of these men's talents, but included them in the class of artisans from which so many of them were drawn. This social gap could only be "bridged by great talent", and the artists themselves, conscious of their gifts and worth as they were, were no less conscious of the necessity for holding a candle to the devil, and treating the only patrons they could expect with the most profound respect.

The efforts of these artisan-artists were stimulated by the sale in fairly large numbers of Dutch paintings in London during the early eighteenth century when some hundred Ruysdaels were auctioned between 1748 and 1750, and it was already possible for Horace Walpole to write and publish his *Anecdotes of painting in England* between 1762 and 1771. There was also a possible market for Dutch paintings in East Anglia where the wealthier merchants had close connexions with Holland, and where the Norwich School may be said to have been founded by Crome at a slightly later date. But the geographical contiguity of the Low Countries and East Anglia was somewhat local in its effect upon the relation between the artists on the one hand and the well-to-do wool merchants and clothiers on the other. Italy was the land of culture where the sons of great landowners, more especially the oldest son, went to complete their education by what came to be called the Grand Tour in the early eighteenth century, though the term has been applied to earlier journeyings by Esther Moir.[13] These continental travels followed a well-established route through France and across the Alps to Milan, Rome, Florence and so on. They did not usually include a visit to Flanders and the Netherlands.

The cost of these journeys, which often lasted for several years, might be very great, no less than thousands of pounds being spent by some of these young men and their tutors, servants and other attendants. There was, too, some degree of orthodoxy about the plan of the tours. The proper people were visited, the proper buildings, churches, relics of ancient times, the correct art exhibitions and collections were seen, and occasionally recorded as by Addison. The consequence was that a system of similar impressions was carried home, and no doubt supplied conversation when social visits were made.

There was a great difference between tourists. The heir was given a much more splendid entourage and a much larger supply of money than the younger sons who spent a year or so in France at a cheaper rate; but the arrangements made and consequently the impressions collected were similar results from a similar experience, however disparate the length of the tour, the number of countries visited, and the splendour of the equipage of the tourist. The final result was the creation of a correct taste in the restricted circles of those who were able to expand their experience in this way. The tourists did not lose any of their inherent patriotism, nor their assurance that their country was the best and finest in the known world. The Grand Tour is one reason for the over emphasis some writers have placed upon the Italian influence on the collection of pictures by the *cognoscenti* of tourism and the resulting influence on English artists.

The tourists were no doubt impressed by the tremendous and appalling landscapes produced by Salvator Rosa, coming as they did into Italy after a hazardous crossing of the Alps. This experience must have made an indelible impression upon their minds, but was to some extent corrected by the less frenetic atmosphere of the paintings of Claude. Admiration of Italy descended from the Renaissance English visitors, and cannot be dismissed as non-existent. The pictures of these artists continued to be admired and collected throughout the eighteenth century, but there were other influences at work. The production of English landscape art was encouraged by the new appearance of the scene created by the progress of the enclosure movement. The artists painted a quieter and more precise delineation of what they saw around them besides toadying to the fashionable taste for picturesque scenes of mountains, fake ruins, tumbling waterfalls, which are in the proper sense created by the grandeur of Alpine scenery, such as never was, depicted by Salvator Rosa and his school. Some indeed preferred imagination to the precision now to be found in the modern tourist's coloured photographs.

Before the production of real landscape painting became the vogue there was in the work of these men something that English pride has believed to be unique in the world, a galaxy of partly topographical, partly landscape artists, who preceded the real departure from convention made by Richard Wilson and others, and more particularly Gainsborough. The men who helped to change topographical records into landscape painting were quite few. Less than half a dozen need be mentioned here. They were: Paul Sandby, 1725–1809; Francis Towne, 1740–1816; Michael Rooker, 1743–1801; Thomas Hearne, 1744–1817; Thomas Malton, 1748–1804. Clearly most of these men did not come to maturity until the next period (arbitrarily selected as that is); but in this list the most distinguished names have so far been omitted. These were Richard Wilson, George Stubbs, and Thomas Gainsborough.

The landscape of England was the most characteristic creation of the British people. Nowhere else have the same factors of soil, climate and history produced anything like it. The rate of change was becoming more rapid as the century progressed. R. H. Mottram declared in 1937 that "the spread of landscape painting", which these three painters had not precisely initiated, but had formed into the mode, "coincided to an extraordinary degree with the change in the outward and visible appearance of the country. Until a hundred years ago this was largely made up of those wide spaces of heath and water, which were the relics of ancient forest, and the already considerable hedgerow or dense plantation, the perennial covering of our English earth". Thus the growth of landscape painting marks and memorialises the eternal conditions of our island life".[14] According to Christopher Hussey some part of the novel appearance of some part of the countryside was related to the needs of the fox-hunting squires and others. The sport was becoming formalised, and its new techniques favoured small coverts, runs across agricultural land and jumping hedges:[15] but this sport cannot have had the same results on the landscape as the needs of the new farming systems, or that is an immediately justifiable opinion.

It is fortunate that the period under discussion is prior to 1760 because the Polite Arts were so little cultivated by the *Natives* of England before that magical date that foreigners, and even many of our own countrymen, imagined that some natural causes prevented the English from becoming masters either in painting or sculpture, an opinion it is impossible to agree with, though its perpetrator was althogether sensible in other respects. Certainly his opinion of Richard Wilson (1714–1782) is just. Wilson has been uniformly praised, called the Father of English Landscape Painting, a man who, though he had visited Rome, wisely avoided any imitation of the pictures of Italian masters, that "wavy line of mountains which bounded the distance in every view", and "the use of dreary inhospitable plains. Nor did he use the mouldering fragments of temples, tombs and aqueducts", which figure both in pictures painted during his lifetime, and indeed much later. On the other hand, Wilson is said to have perpetuated a feeling for landscape akin to Claude; and Turberville assessed him as the greatest eighteenth-century landscape painter, which is surely an exaggeration.[16] He is not free from some of the elements of the picturesque which obsessed aesthetes and artists alike a few decades later. Rothenstein mentioned particularly Wilson's *On Hounslow Heath* (Pl. 8), c. 1770, a waste with a pool and an island, some figures and cattle, sombre in colour and with a darkling sky, somewhat of the pattern of landscape remarked by Mottram. Ruskin pronounced Wilson a sincere landsscape artist. Amidst so much counsel the emergent is that Richard Wilson was one of the most outstanding of the precursors of the great age of English landscape art, and that he saw Hounslow Heath as a kind of desolation – which it must have been – "a touch of that impossible realism . . . the fact (which none but the ignorant dispute) that no such combination of scenery exists in nature as the painter of genius may produce. No such paradises are to be found in reality as have glowed on the canvasses of Claude . . ." This may perhaps have been the origin of Wilde's dictum that nobody had appreciated the mists on the Thames until Whistler painted them.

Wilson was not singular, though outstanding. There were other artists contemporary with him, if not of his stature. Thomas Smith of Derby was one of them, a landscapist, the father of an artist, John Raphael Smith, the brother of an artist, as Julia Frankau put it.[17]

The bucolic English were becoming sophisticated, but those who were formed the small and aristocratic minority. They patronised art but did not practice it, and it might reasonably be enquired how art in the form of painting could flourish otherwise? The esoteric too were changing. Edmund Burke published his *Philosophical Enquiry into the Origin of our Ideas of the Sublime and Beautiful* in 1753, and in December of that year the pugnacious Hogarth his *Anatomy of Beauty.* He introduced moral fables and the conversation piece, a venue much loved by Zoffany, Philippe Mercier, Laroon and Joseph Highmore.

There had been few or no artists of these tastes before them. The same could not be said of George Stubbs (1724–1806). There had been many horse painters before him, and artists of racing scenes, but Stubbs was in a class by himself. There were many animal painters after him during the following two centuries: Sawrey Gilpin born 1723; Benjamin Marshall, 1767–1835; John James Chalon, 1778–1854; James Ward, 1769–1859; the Herrings, and most recent of all, Munnings; but that is not precisely the point here. Stubbs was a realist. He was obsessed, if that is the right word, by anatomy, human and equine, and his extraordinary feats of strength and stamina when making the dissections that preceded the publication of his *Anatomy of a Horse* have become so well known that it would be useless to repeat them here.[18] His reputation has recently led both to the reprinting of his major and original work, and to the publication of more than one biographical appreciation.[19] He did not paint landscapes in the special sense of the word in mind here, but he did paint pictures of rural occupations, the Haymakers, Hay cutting, Reapers, the Labourers (Pl. 9), but these are no more landscape than are those of people in the saddle or in a horse-drawn carriage of one sort or another. The farming processes are action pictures, and are a lead into landscape such as Gainsborough delighted in (Pl. 10). His painting of the vast and renowned Lincolnshire ox must not be forgotten (Pl. 11). Pictures of racehorses in their environment, not in the stables, but in the enclosing scenery, must have led to the painting of scenery for its own sake. It could be seen and appreciated in the horse pictures, and came to be appreciated without the intervention of the animals.

Landscape for Stubbs was an adjunct to his racehorses, his riders, his carriage folk, and the workers in the fields; for Gainsborough the countryside made the picture, and human activities were incidental. He was, even more than Richard Wilson, even more than any of the topographical artists, percipient of the countryside as a pictorial scene not requiring human drama or occupation as a focal point. Of course this is not true of his portraits, more particularly the famous Andrews picture (Pl. 12). The two young people in this picture are looking over a rural landscape that has changed but little in the two centuries since the picture was painted. Gainsborough loved landscape, but he loved the good things of this life still more, and was therefore forced to become primarily a portrait painter, much as professedly he would have preferred to devote himself to the changing countryside. His pictures are well known, and Cornard Wood (Pl. 13) and the Market Cart (Pl. 14) show bits of the countryside where he was born, and where now the proud citizens of his birthplace have formed a cultural and artistic centre at Gainsborough's House, his reputed birthplace – about which there is some dispute.

By the end of this arbitrarily adopted period – many of the outstanding painters were born in it and lived long after – the change in farming systems was accelerating, and it was a far cry from the time when Lowell, writing on

Pope, could speak of Bolingbroke raising delusive turnips on his farm. Turnips by this time (1760) were far from delusive. They had become the lynchpin which, with the grasses, clover, rye grass, etc., were making all rural people who cultivated them more prosperous, except, of course, the labourers, and impressing on the hierarchy of the countryside all the deceptive and apparent permanence which is one of the most evanescent of all human experience.

John Mills, writing in 1762, protested that "The appellations rude and rustic were never given to really useful occupations till luxury had debauched mankind and a frivolous pertness usurped the place of sense and manners". This is less than just, and would certainly have been resented by the country gentlemen who set the standard of behaviour and most properly of improved farming systems. The leading lights of this social class certainly spent some part of the year in London, but most of their time was passed on their estates, where they practised field sports, perhaps to as great excess as they practised drinking, but also gave their benediction to the innovations being made under their supervision upon their farms. They enjoyed the mildly social pleasures of rural living and the society of their equals, even of their social inferiors, who all or most of them knew their place, however dissatisfied with it they may have been. In the words of Christopher Hussey, "the squire lived like a lord, the yeoman in the manner of a squire (when funds allowed), and the villagers to the best of their ability (it must have been little enough if it existed at all) as yeomen".[20]

The accession of George III marked what might not inappropriately be called the end of the beginning. The new farming was well under way, and the physical appearence of at least some of the rural scene was being changed by the needs of the novel features of farming, the increasing wealth of the great landowners was finding expression in the building of magnificent palaces set in a framework of artificial landscape gardening, and the painters who were being called upon to record these things were bringing a new view to the novel scene. All this had to be explained by new, or novel, formulae for the appreciation of the sublime and beautiful, which were to be absorbed in theories of the picturesque in the next age. And the first exhibition of pictures, etc., of the newly formed Royal Academy was held in 1760 in the rooms of the Society for the Encouragement of Arts, etc., that Society itself awarding prizes for artistic work in the same year.

NOTES

1. Cf. J. T. Krause in *Land, Labour and Population in the Industrial Revolution*, ed. by E. L. Jones and G. E. Mingay, pp. 187–205.
2. See G. Melvyn Howe, *Man, environment and disease in Britain. A medical geography*

of Britain through the ages, David and Charles, 1972.
3. D. C. Coleman, "Gentlemen and Players", *Economic History Review*, 2nd series, XXVI, i, Feb. 1973.
4. Francis Bacon, "Of Gardens", *Essays*, London, 1902, p. 129.
5. E. Malins, *English landscaping and literature, 1660–1840*, Oxford University Press, 1966, pp. 28–9.
6. Dorothy Stroud, *Capability Brown*, Country Life, 1950; Edward Hyams, *Capability Brown and Humphrey Repton*, Dent, 1971: cf. Malins, cited n. 5, pp. 28–9, 112–6.
7. Dorothy Stroud, cited note 6.
8. Leslie Stephen, *English literature and society in the 18th century*, Duckworth, 1910, p. 175.
9. James Thomson, *Poetical Works*, Oxford, 1908, pp. 4–5.
10. H. Morley, *First sketch of English literature*, 13th edition, 1887.
11. See above, notes 5 and 6.
12. Sterne's *Tristram Shandy* for the details.
13. Esther Moir, *The discovery of Britain. The English tourists, 1540–1840*, Routledge, 1964.
14. R. H. Mottram, *Old England illustrated by English paintings of the 18th and early 19th centuries*, Studio, 1937.
15. C. Hussey, *The Picturesque. Studies of a point of view*. Putnam's Sons, New York, 1927.
16. A. S. Turberville, *English men and manners in the 19th century*, Clarendon Press, 1929.
17. J. Frankau, *An 18th century artist and engraver, John Raphael Smith: his life and works*, Macmillan, 1902.
18. C. A. Parker, *Mr. Stubbs, the horse painter*, J. A. Allen and Co., 1971.
19. C. A. Parker, cited note 18, and B. Taylor, *Stubbs*, Phaidon, 1971.
20. C. Hussey, *British country life in art*, Country Life, 1937.

III

The Great Age of Improved Farming and Landscape Art, 1760–1815

A large proportion of our ancestors, although inclined, as we all are, to accept the conventions of the time, more especially in behaviour, were only too willing to break away from contemporary working habits and customs. They did this so effectively that they changed the appearance of the landscape of some 2½ million acres of common field with some waste, and of an additional ¾ million acres of waste land between 1761 and 1801.

The acres enclosed and divided up into fields separated by hedge and ditch, by post and rails, by dry stone walls and so on, sounds and is a large area, but it was not contiguous, and some of it was separated from other parts by those spaces of uncultivated or relatively uncultivated land remarked upon in a rather exaggerated way by R. H. Mottram.[1] The application of the new agricultural methods to these newly enclosed lands was not always immediate: the actual physical redistribution and enclosure was sometimes delayed for an undesirably long time, but both things were finally accomplished to the great advantage of the people concerned. The increased output of corn and meat, dairy and poultry products was very important for the sustenance of the more numerous population that was becoming concentrated in the industrial towns. Even so, the home production of cereals did not wholly meet the demand after about 1770.

Five other towns besides London had reached sizeable dimensions by the time of the first census. Manchester, Bristol, Birmingham, Liverpool, Leeds were then housing over 50,000 population each, but these people were mainly workers in the new factories, or their employers, and the towns themselves were the products of an almost insane economy in building, small houses crowded in uncomfortably close rows put up to house the helots of the nascent industrial system, just as lacking in sanitary conditions as the mansions of the nobility. This certainly changed the appearance of the countryside, mainly in the north of England, but did not give it beauty, creating something it is true, but something rather disastrous in many ways, turning what had perhaps been a blank and savage scene into one of the most forbidding aspect. This was something quite different from the current pattern of building for the better-off people. In the eighteenth century new methods of work were making agriculture and other industries more productive, and those who benefitted, directly or indirectly, loved to build. The aristocracy and the newly-rich employed architects to raise mansions: squires ordered gracious country houses with pillared Palladian porticoes, lesser squires and business and professional men put up small Georgian residences in brick or stone.

The cottages in which the rural labourers lived were by no means so pretentious. They were not equal to the two up and two down dwellings in the rows of the new towns, or in the mining valleys of South Wales. Usually, or often, the cottage was a one room affair with a lean-to shed, and though it looked picturesque in the eyes of the painters and on their canvases, must have been crowded, noisome and insanitary, drains being non-existent and seepage into the well of undesirable and disease laden matters common. So it did not matter very much to the human beings living in such fragrant amenities if they were forcibly removed in order to improve the landowner's outlook, or were turned out without other provision, as at West Sheen where 18 dwellings were demolished in 1769. The site was made into a lawn, and added to George III's enclosures, later being made part of Kew Gardens. Some of the few new villages, like Milton Abbas and Nuneham Courtenay, had much better cottages with several rooms, and were a credit to the "cruel" landowners who had them built. In the greatest contrast to these humble rustic homes, and to those in the new and growing industrial towns, were the elegant structures in imitation of those at Bath which were erected in the Spas, Cheltenham, Scarborough and Buxton, where the rich, the not so rich, and their serviteurs and parasites visited for pleasure combined with drinking the waters for sanitative reasons.

A glimpse of the obvious is that these improvements, which all the changes were not, were made possible by the continued prosperity of the country, the beginnings of the industrial and the progress of the agricultural revolution. It is quite impossible to give any arithmetical measure of the output of farming, whether crops or livestock, fruit or hops, but it can safely be assumed that, with the enclosures making it easier to protect the growing crops, and keeping the livestock from wandering aimlessly over the wastes, there was some interest in output in relation to area in many places. This is made clear by the exploits of individual farmers, both in the arable fields and in breeding, something that is described in every textbook of farming history, and in a great variety of papers in learned periodicals.

The last forty years of the eighteenth century and the first decades of the nineteenth witnessed the birth and rise to fame of a galaxy of great men, great men who made their paths to glory along many different routes. Because it is fundamental to every other human activity it was the great men of the farming world who laid the strong foundations of what we are pleased to regard as progress, though the word needs a more precise definition than its present vague application to all sorts of developments. In every other sphere of human activity there were similar clever men, men of genius if that is not too high a rating; but the agricultural pioneer, propagandists and pundits laid the foundations on which all else must be built.

Every schoolboy knows the names, if no more, of the outstanding men who energised the rural people in the late eighteenth century. It is only necessary here to name them once again. Arthur Young was the foremost propagandist, who publicised not only the new systems followed here and there in the countryside, but also the novel processes that unknown farmers had worked out and practised in the isolation of the villages. By making these things known through the publication of his *Tours* he stimulated yet more than current circumstances the publication of farming textbooks, some of which contained the greatest nonsense, but others the most sterling common-sense of that day. William Marshall was a literary competitor at a later date, and between these two and the mass of current farming books an almost complete idea of the practices of the day can be obtained. By making his *Tours* Young secured a remarkable collection of information, and this was heavily reinforced when the Board of Agriculture was set up with Sir John Sinclair as President. It immediately set about making a complete survey of the kingdom's farming, the results being enshrined in what are known as the *County Reports*.[2] This was something the Royal Society had tried to do in Charles II's reign, but had found the task impossible, that is to say if the records have not been lost.

Mention of Sir John Sinclair makes the work of practical men spring to mind. The fodder supplies resulting from the adoption of the new systems made it possible for enterprising men to breed and feed larger and more productive animals. The name of Robert Bakewell of Dishley Grange is another that is found in every textbook. He was preceded by Sir Thomas Gresley of Drakelow, Derbyshire, and Webster of Canley, and followed by other famous men like the Colling Brothers, Thomas Bates of Kirklevington, and George Culley. There were many others, but there is nothing so tedious and almost uninformative as a list of names.

Many of these breeders wanted portraits of their outstanding animals painted. These pictures were not, of course, landscape in the proper sense, but in those where the animals do not occupy too much of the foreground there is landscape in the background just as there was in the earlier pictures of religious episodes and classical legends. Animal painting is a subject that has recently been treated in some detail by Basil Taylor.[3]

The breeding of fine horses, breeds of cattle, sheep and pigs is a business that lends itself to pictorial art rather more immediately than the procession of the seasons as arable and fodder crops grow in the fields. Nevertheless the seasonal occupations have lent themselves to pictorial representation, often somewhat crudely, for more than a thousand years. The protagonists of the agricultural revolution in the arable fields of the country had only the wider areas under crop divided by the new (in some places) and spreading hedgerows to offer as subjects to the artist. The men who worked their fields in the way of the new husbandry were often great landowners, and here again their names are familiar.

The end of the eighteenth century resounds with the names of the famous landowners, whose labours not only made the country what it remained for a century, but also enriched themselves beyond anything even their acquisitive instincts and desires for conspicuous expenditure could have believed possible. Here again the most familiar names are those which figure in the most elementary of economic history textbooks, but there are some who are not quite so familiar, possibly being less flamboyant personalities. Coke of Norfolk and the Holkham Sheep Shearings need no bouquet here; nor does the Duke of Bedford who held similar meetings at Woburn Abbey. The Royal Agricultural Society of England has a remarkable painting of one of these meetings, which, though not landscape, seems to need mention here if only because of the large number of portraits of agricultural celebrities it includes, with their names given no doubt in the very present hope of a suitable solatium. Other great landowners played their part in the reclamation of land, or in admirable attempts at a sort of pseudo-philanthropy based on setting the poor at work, a process that was carried out by Curwen of Workington, Cumberland, to the great benefit of the local helots. Other landowners, too numerous to mention here, did something of the same sort, and all over the country these people formed themselves into associations in order to promote the new husbandry, and to stimulate the more and more widespread introduction of the novel and more profitable ways of working.

The first of these in London was what is now known as the Royal Society of Arts, formed at a meeting held in 1754 at Rawthmell's Coffee House in Henrietta Street, Covent Garden, and attended by noblemen, clergy, gentlemen and merchants. An odd event was that when the Secretaryship became vacant in 1760, Oliver Goldsmith applied for the job, but Garrick's refusal to support him – because Goldsmith had criticised him adversely – caused his application to be refused. This Society offered premiums for all sorts of farming experiments, and was also interested in stimulating the arts of painting and drawing. It had some part, minuscule perhaps, in the formation of the Royal Academy. It is not easy to overestimate the work of a society of this kind which grouped together so many eminent men, and men of high position in the country, but the claim made that the cultivation of the swede, the turnip and the mangel wurzel were introduced to our fields by this society is quite untenable.

The Society for the Encouragement of Arts was not specifically an institution interested in farming, though the occupation played a large part in its initial activities, offering premiums, medals and so on. Two other societies were formed to promote agriculture. They were the Bath and West of England Society, which was founded in 1777 and continues to flourish as does the Highland and Agricultural Society established in 1784. It can hardly be said to be a successor to the (Scottish) Society of Improvers in Agriculture founded in 1723, which collapsed in 1745 though it had similar aims. The Smithfield Club was started in 1798, but five years before this the semi-official Board of Agriculture was set up. As already said, this Board is mainly remembered for its series of County Reports on the current situation of farming, although it had several other objects, making enquiries into this and that, and publishing the results in a series of *Communications*. William Marshall, who made a one-man survey of the nation's farming and wrote a condensed version or summary of the County Reports, was another agriculturist of equal stature.[4]

These were men who not only influenced farming by their writings but farmed themselves, Young not too successfully. There were famous practical men as well. Foremost in fame is perhaps Coke of Holkham, followed by the Duke of Bedford. Both of these held some of the most outstanding of the contemporary agricultural shows, then known as sheep-shearings, and these privately organised exhibitions were nearly as all-embracing as the modern shows run by Societies. Besides these two there were many lesser lights whose farms have not been so emphasised in the textbooks. Among these were Sir Digby Legard, who was responsible for massive reclamation in the Wolds of Yorkshire, Lord Raby of the same county, who was by way of being a friend of Arthur Young, and slightly later, John Christian Curwen of Cumberland who made some efforts to improve the lot of his tenants and workers. A few landowners like these, scattered about the countryside, were able to exert, not exercise, weighty influence on the proceedings of their tenants, if only through the punitive clauses in their leases. And, as readers of Arthur Young's *Tours* and his periodical, *Annals of Agriculture*, cannot fail to realise, such men, owners of the land and almost of its populace, were numerous.

So the agricultural revolution was eagerly urged on, both by the theorists and the practical men, to the financial advantage of both, and to the growing prosperity of the nation. There were yet other famous men in the undertakings that had served the farmers. The day of the village blacksmith and carpenter had not yet sunk into its evening, but, and this is an emphatic but, the blacksmith was developing into the engineer, though the carpenter and wheelwright continued to exercise their skill throughout the horse age even into the twentieth century when horses vanished from the land.

A theory of the plough had been studied by Capt. Walter Blith in the time of the Commonwealth, and his work was followed by that of the other theorists in the latter part of the eighteenth century. Jethro Tull had produced the first usable seed drill late in the seventeenth century, and had many apostles during the eighteenth and early nineteenth, the most long-lived of whom were Smyths of Peasenhall. Robert Ransome started his plough works at Ipswich in 1783, which has developed into one of the most famous agricultural implement factories in the country. Threshing machines were to be obtained, and many minor, but useful, appliances of a semi-mechanical sort were designed for use in the barn. The men who designed these things and made them were not all drawn from what would now be called the working class, a denomination modern members of which would probably resent instead of being proud of it.[5]

One was James Sharp, who produced the first hand operated chaff cutter, and who has returned to fame as a pioneer of rolling carts. A recent article in *Country Life* gives him more social status than I accorded him in an earlier work. [6] This man, like so many of the other innovators, was not one of the depressed and oppressed manual workers. He was a descendant of a distinguished family, who has been called a rough ironmaster, he and his wife being nicknamed Vulcan and Venus. "But more he was navigator, engineer, and probably the first man to make and advertise farm implements on a national scale". He was of sufficient social status as to have his family painted as a group by Zoffany (Pl. 15).

All this ferment of activity created the need for more mobility. The concentration camps of the industrial revolution massed a non-food-producing population in the ghettoes of the new towns. These people must be supplied with at least a modicum of food, and it had to be carried to them. London was a vast magnet for all sorts of supplies as well as being at that time the major centre of the nation's commerce. East Anglia and the Bristol Channel were on the outskirts of this, but were transit points if no more. So the roads must be improved as well for the vehicles as for the four-footed beasts that were so large a proportion of the traffic.

This is not the place to tell the story of the highway, first because it has been done so much better by other people, [7] and second because it is on the fringe of the present subject, yet it must be hinted at if no more. The novel developments all involved transport, the carriage trade, the coaches, the heavy wagons, the animals and even the poultry upon their own feet. The markets for agricultural goods were fast becoming centralised. London had always been the lodestone, but towards the end of the eighteenth century other massive centres of consumer population were forming, and these centres themselves were producing heavy goods, ironware, ceramics, fabrics, etc., that were not consumed in the producing centres, or only to a minimal degree, and must be sold in distant markets to provide the financial resources that was their lifeblood.

Arthur Young's complaints about the foundrous roads he found all over the country when he made his *Tours* in 1769 and the following years leave no doubt about the difficulties, not to say dangers, confronting a traveller then. These are amply confirmed by other travellers, and by the fiction writers of the peripatetic novels of the day. The travels of *Joseph Andrews* and his mentor Parson Adams, the vagaries of *Tom Jones*, and others, who at once spring to mind, leave no doubt about the hardships of the journeys involved. The nobility went accompanied by baggage waggons and armed outriders guarding the large sum of money in golden guineas and real silver coins that had to be carried along to pay the massive expenses of such a journey. It is at once apparent that travel was restricted to the rich or to the businessman who must endure these hardships of necessity. As to the minor farmers and "the peasants of old time", as George Eliot put it in *Silas Marner*, "the world outside their own direct experience was a region of vagueness and mystery". On December 11th, 1786, William Cowper wrote a letter to Lady Hesketh in which he said that Mr. Chester had paid him a morning visit, but had "begged him not to return it (the visit) with a punctilious alacrity in this wet and dirty season: – an allowance for which I am obliged to him since we now live five miles asunder and I never ride, it does not at present occur to me by what means I could possibly get at him", a sufficient commentary on the winter state of the country lanes. They were little if any better in the early years of the present century, as my own experience can substantiate.

Transit became a major problem with the rising volume of traffic of all kinds, and one solution was the setting up of turnpike trusts to take the place of the earlier compulsory labour of each parish, which, like all compulsory labour, was virtually ineffective. Many of the turnpike trusts were little better, but the story has been told so often that it could not be conscientiously repeated here. With the need comes the men, and great men appeared to undertake the task of road making. The outstanding names are, Blind Jack of Knaresborough, Telford and Macadam, household words to all students of economic history. Doubtless their example, extensive as it was, stimulated local road makers who must have helped to make the system viable. One thing about all this movement that touches upon the present theme is that all these travellers, whether men of means, commercial men, or those like Parson Adams making their way to fame and fortune in London, must have observed the surrounding scene as they made their slow progress from one place to another. They could not help but look at the landscape, and there can be little doubt that although they might themselves be obliged to have resort to some village alehouse, they looked upon the noble mansions surrounded by their landscaped gardens with the same awe and admiration as their motoring descendants. They could not avoid "views", however much they may have disliked, even feared, the physical landscape through which they travelled.

When benighted, the terrors and hardships of the traveller were magnified. An emphasis of this is the precaution of the Lunar Society of Birmingham to hold their meetings only in the full of the moon. This society, which collapsed when the French Revolutionary Wars broke out, had a distinguished membership, including Joseph Priestley and Erasmus Darwin. Another very learned provincial organisation was the Manchester Philosophical Society, which became very distinguished and achieved an international reputation. There were many lesser lights, and there were social and literary gatherings in many of the Assembly Rooms which graced centres like Bury St. Edmunds or Lewes, where the gentry and city fathers met and mingled.

All the difficulty in road transport and travelling had time out of mind been mitigated by coastal shipping, barges and so on, and wherever possible along the routes of navigable rivers. The benefit of using the latter had been emphasised by Taylor the Water Poet in the early seventeenth century, and proposals for man-made waterways had been made by Rowland Vaughan at about the same time. These hints of ways and means took a long time to come to fruition, but by the 1760s they had been given by the Duke of Bridgewater, who was responsible for the construction of the Manchester Ship Canal. Only a few decades passed before there was a network of canals, some connecting rivers, like the Thames and Kennet, others merely joining one another. Some quite extraordinary enterprises were undertaken as always happens when a new idea is adopted too enthusiastically, possibly by the uninstructed fortune seeker. Water transport of this kind made it easier and cheaper, possibly more speedy, to carry supplies of grain and other heavy foodstuffs to more numerous centres, more densely populated, of consumption. Raw materials, ironwork and coal went along the canals with a good deal less difficulty than they could be hauled along the still inadequate roads. These, like the hedges and other boundaries of the newly enclosed fields, had a marked effect upon the appearance of the countryside, just as railway building did in the nineteenth century, and arterial roads for the convenience of motor traffic are doing today. The new scene made an immediate appeal to one painter, and seems to have had a profound effect upon his work. This was Constable, and his observation and record of the changes has educated the eyes of many a non-artist to seeing the beauty of the new scene. Water scenes, rivers and canals are very significant in his art, but this is to leap ahead.

All these agricultural, industrial and commercial activities were carried on in a forbidding framework. It was an age of conflict and blood sports. War was endemic, if not always official, but the business of privateers (pirates) and adventurers. The populace from high to low loved conflict between nations, all foreigners belonging to a sub-human species, and fighting was a sport not only between nations, but between individuals, pugilists, wrestlers and so on, and the irrational animals were recruited to provide sport, bear and bull baiting, cock fighting, dog and rat fights and other elevating pursuits upon which wagers could be placed. The artist of the time was not immune from this spirit. The artists painted pictures of animal conflicts of one kind and another. George Stubbs was a past master at this kind of thing; James Ward was another, and they are not isolated examples. It was an odd concatenation that this preoccupation could be combined with the most delicate appreciation of a landscape, and its meticulous, inspiring reproduction by the same artists in another mood. This was a mood stimulated by the taste of the landowners. It need hardly be said that these tastes were modified, if not created, by the circumstance that the leading and most productive farms supplied the noble and gentle landowners (who used spring guns and mantraps in their coverts) with even bigger incomes than of yore. This accretion of revenue enabled them to indulge their taste for what was becoming known as the picturesque, by landscape gardening, which was not exactly a new hobby, but when combined with the *cottage ornée,* could perhaps be thought to be so. There is no doubt that the men who were keen on these things were also keen on buying pictures delineating the subjects in which they delighted, stately houses, beautiful gardens, picturesque cottages and flourishing fields and farms.

The changes that men were making in the living scenery in which most of them spent their lives in the late eighteenth century must have made an impression upon their senses, vision in particular. They must have seen the landscape that they were so completely making over, but it can justifiably be doubted whether they saw it until the artists had, so to speak, pointed the moral in their productions. Dare it be said that the picturesque suddenly became a catchword. It would be impossible to deny that it was already known – the antiquity of the Italian word *pittoresco* attests to that – but like other things of a respectable age this term became commonplace, and many people undertook to define it in the last few decades of the eighteenth century.

William Gilpin was one of the high priests of the picturesque, and he was a persistent tourist in those days when touring was a far more adventurous, sometimes perilous, hobby than it has become today. He made tours in search of picturesque scenes in many, if not most, parts of the kingdom between 1769 and 1808, and he published a volume of description after each journey. For him it was the grandeur of a scene that was impressive. He loved towering rocks and tumbling waterfalls, and those were some of the ingredients used by the more conventional artists. Some forms of the picturesque were but little less imaginative than the earlier cult of the horrific by such artists as Salvator Rosa. Forbidding mountains with their rock faces, plunging waterfalls (such as Gilpin liked), all the kinds of terrain that are nocuous to man were the themes the later disciples of the picturesque loved, such as can be pin-pointed in their paintings of the worst declivities of mountains like Snowdon or the Lake District, which they humanised a trifle by introducing domestic cattle grazing in the vales or sheep and their shepherds with dogs in the higher moorlands or summer feeding places.

Gilpin was not singular in making his tours in different parts of England and Wales. The foreign travels of the scions of nobility who made the Grand Tour, which they continued to do, were not exactly imitated by the growing number of explorers who made their journeys through unknown England and even more unknown Wales. Even the hypochondriacal Samuel Johnson was bold enough to make a difficult and protracted journey to the Hebrides.

Some people had made tours in England in the seventeenth century; but the earliest practically complete visit to all parts of the country was made by Celia Fiennes at the end of that century. [8] The topographers, who made what would possibly be called economic geographical surveys of the country as a whole, were not concerned with the landscape as Celia was, not excluding the difficulties encountered in her travels. It cannot be said that Celia created a fashion, but she was followed, mainly only in particular sections of the country by a fairly large number of our intrepid ancestors throughout the following century.

This is not the place to follow these tours *in extenso*. That has been superlatively done by Esther Moir but a few years ago, and a little before that, but in less detail and with a different slant, by Rosamund Bayne Powell; but the famous Arthur Young's *Tours* must be mentioned, and the farmer's and landowner's journeys described in his *Annals of Agriculture*, not to speak of the *Rural Economy* series written by William Marshall, although these have been noted above.

These men were travelling with a purpose quite different from the objectives of the ordinary tourist, if in fact there was such a person: but they did not fail to remark upon the beauties or terrors of the landscape, and they did not escape the unofficial charges made by the housekeepers and other servants in the great houses they visited. In those days the owners did not need the vails (tips) given to the servants, or demanded by them. One housekeeper, for instance, demanded five shillings when offered half a crown. The amount she asked was about five days' wages for a farm labourer then. Some of the landowners had a menagerie, one at least including lions, for the delectation of themselves and their inquisitive visitors. Others had catalogues prepared, to simplify the tourists' tasks. As Esther Moir put it, "the tourists descended like vultures on their prey sweeping through the house and gardens relentlessly determined to see everything", [9] and to pass judgement in their records upon the taste of the owners.

Arthur Young, whose tours were made with such a different object, fortunately did not escape this temptation. His main interest, it is needless to say, was the current practice of the farmers in the district in the places he visited; but he was not immune to the precious things in the mansions of the nobility, nor to the landscape through which he travelled. A few examples will suffice. He did not like Castle Howard, though it is really very impressive, but the park and grounds at Wentworth Woodhouse, where he was treated with kindness, received favourable comment. Indeed he was as interested as more mundane tourists to inspect newly built ruins, grottoes, what have you, of the prevailing fashion. At Barnard Castle he found "a continued landscape sufficient to captivate the most languid viewer", and this is by no means an isolated comment. He hated the desolation of the wilderness of Salisbury Plain, a sentiment he certainly would have repeated could he have seen it today.[10] "Hedges and fences were the symbols of the new England, bringing the end of the sweeping open fields of the medieval landscape, and heralding changes in farming practice and technique" of which Young was an admiring but caustic critic. William Marshall, too, had moments when he recorded his impressions of scenery, both man-made and what may be called wild nature, of which there was still plenty in those times.[11]

The picturesque was the form created by some painters who worked along what might be called traditional lines. The landscape artist who painted the setting of the agricultural revolution depicted very different, and more homely (if that is the right word) views. Many of these were man-made during enclosure and redistribution of the land previously farmed in open field, but there were other and more conscious attempts to make the surroundings of the magnificent mansions of the nobility and gentry conform to the ideas comprised in the meaning applied to the word picturesque. This was an elaboration of the landscape gardens – of many acres – that Kent and "Capability" Brown had laid out. Naturally this process did not come to an abrupt end at a given date, and towards the end of the century a controversy raged between Sir Uvedale Price, Humphrey Repton, Richard Payne Knight and others. These people were eloquent, and propounded their theories of the picturesque, the sublime and beautiful, landscape painting and landscape gardening in numerous works. These enthusiasts probably worked, no doubt unconsciously, on the basis that print, however controversial it may be, is more convincing than ephemeral verbal arguments that may have circled about the subject; but the printed word, alas, has only the appearance of authority, and is just as susceptible to criticism as the resounding vociferations of wordy discussion.

The controversy went on for a couple of decades, and during the same period the beautifying of estates on picturesque principles continued. The landowners in the less mountainous and more cultivated areas began to adorn their landscaped estates with picturesque cottages, some of which remain in occupation to this day, and are admired by the motoring tourists as oddities, as indeed they were and are. The craze lasted well into the nineteenth century, and enterprising architects produced books of designs for such things. The older cottages were picturesque enough without amendment, though repair and even rebuilding might have improved the lot of the tenants, who probably did not care much about the appearance of their cramped insanitary and often ruinous homes. To build a *cottage ornée* in the grounds was quite a different matter, and added something to the already pleasing aspect from the windows or terrace of the mansion. Action of this kind made the contrast between the comparative comfort of the new cottages and the bleakness of life in the older ones more plain, though the sanitary condition of the new dwellings was often no better than that of the old. Appearance was all. The roseate hues of the new buildings under the eyes of the great contrasted with the bitter gloom of the old. Some odd people, and there were many odd people then, developed idiosyncracies from living in a small, slow-moving society. Today it would certainly be a curious thing to pay a hermit to live in the grounds, and sit at the door with a book in his hand, which he probably could not read, when visitors were about – a hint to the modern descendants of these gifted showmen.

One of the oddest results of the love of the picturesque was the proliferation of "White Horses" carved in the side of the chalk downs, a bit like the regimental badges of more modern days. These could create a difficulty for the archaeologists, but the story has been told by Ralph Whitlock. Some of them are prehistoric; others were carved in the late eighteenth century, and later still in the nineteenth. Ralph Whitlock has the gist of the matter, and nothing can be better than to quote him, especially as his remarks are coincident with the opinions expressed here.[12]

"The phenomenon of the crop of hillside white horses in the eighteenth and nineteenth centuries can probably be explained (certainly would be a better word than probably) by the interest in landscape gardening awakened by the enclosure acts. These gave great sections of the countryside formerly owned communally by the parishes into the hands of great landowners, who began to vie with each other in enhancing the splendour of their domains. To their activities we owe many of the familiar features of the rural scene, such as the layout of large arable fields, the shelter belts, coppices and hill top clumps of trees, the follies, the ha has, the fake antiquities and artificial grottoes", all of which has been said here, but not that the "White Horses were a further example of their exuberance".

One or two of the examples mentioned by Whitlock are the Westbury white horse, which was laid out in 1778,[13] the Cherhill example carved in 1780 by Dr. Alsop,[14] and the Alton Barnes horse not far away, said to have been designed in 1812 for £20.[15] Others are later still, but the one at Osmington, near Weymouth, which Whitlock thinks was carved in 1815 to celebrate Waterloo, was always said by local people in the early years of the present century to have been carved as an emblem of George III's visits to the town.

This was carrying the picturesque to an extreme, but it helped to make people conscious of the scenes in which they lived. Basil Taylor has made the remark that the artists followed a taste that was cultivated in the second decade of the eighteenth century, and that they produced their landscape paintings in response to the Englishman's discovery of his scenery, and the subsequent cult of the picturesque. Landscape became first an acceptable theme, next a fashionable one. Christopher Hussey declared with some truth that "an appreciation of scenery is a very late acquisition in the development of the western mind". [16] Some exception could be taken to Taylor's pronouncement, because a decisive answer to the question which came first, the artists' appreciation of the pictorial value of the landscape and his offering pictorial descriptions of it, as Gainsborough, Wilson, Stubbs and others were already doing, or whether they produced their pictures in response to a demand, is a question to which there is no answer. Part of a reply might be that Wilson suffered much hardship, Gainsborough was obliged to turn to portrait painting, and Stubbs to animal portraits and gruesome pictures of animals fighting. Sir Joshua Reynolds did not approve of these "fancy" pictures. But whether the demand or the supply came first is possibly not so important as that the landowners were becoming richer and therefore were able to buy the pictures. Possibly the artists did not make a conscious change in the pattern, but rather proceeded to produce goods that would sell. In other words they set out to meet the requirements of what would now be called a developing market, then a market of prideful customers wanting to celebrate their own astucity in being in the forefront of the most advanced farming of the day. The men who flourished began to look at the scenery they created with pleasure, a pleasure that was not shared by the helots of the system. For them, as for their forefathers in the days of the *Promptorium Parvulorum*, and even before that date, the fields were not a scene to be looked at for aesthetic enjoyment, but the environment of their daily work, hard and exacting work in the most searching conditions, but in which, like other skills, it was possible to take pride in the result of their educated labours, though their learning was a banausic, not a literary accomplishment.

These labours were the subject of pastoral poetry no less, probably more, in the eighteenth century than before. Thomson, Dyer and some others have appeared earlier in these pages, but these writers paid little or no attention to the real way of life of the romanticised peasant and worker. Oliver Goldsmith made their plight known in "The Deserted Village". Two other poets, not in the proper sense members of the school that raised the romantic revival at the end of the century, but who were closely acquainted with the life of the poor from personal experience, made this the substance of their work. The masculine forthrightness of Crabbe would leave no doubt in the minds of his readers that there was much suffering amongst the servile helot class. The only wonder is that his poems became so popular. Rags are romantic to the well clothed, but not to their wearers.

Andrew Young suggested a more recondite reason for the manner and moral of Crabbe's poem.[17] It was produced as a contrast to Goldsmith's, but this is a piece of speculative psychology for which there is no substantial foundation. He is not so far off the mark in proclaiming that "The Village" being realistic is something new, a successful novelty encouraged by that indeterminate called the spirit of the times.

Bloomfield was of a very different cast. His poems are the acme of mawkish sentimentimentality, and can be compared with the virtuous nostalgia of so many modern writers about the rural scene. Gray was more robust, but his *Elegy*, emphasising the weariness of the ploughman at the end of the day, is tinged with sadness at the evanescence of the human lot. There were some other villagers who became poets, such as Stephen Duck, though they did not achieve quite the same measure of contemporary notice as Crabbe and Bloomfield, but the rural scene, rustic life, and its imaginary satisfactions are a never ending topic. Bloomfield's villagers have some resemblance to the yokels who figure in George Morland's pictures, who look to the modern eye so comfortable and happy (Pl. 16), but though rustic in content Morland's pictures are not true landscape. He was late in the period.

It is generally agreed, or if not agreed at least repeated by most of the modern writers about landscape painting, that its development in England took place in the second half of the eighteenth century; but this is not so accurate as it would appear at first glance. There were painters of what might be called pseudo-landscape before that magical division, but though influential they were not numerous. The names of Wilson, Gainsborough and Sawray Gilpin at once spring to mind, but in point of fact, if there is such a thing, it is more particularly in the fourth quarter of the century that this special form of art was pursued by the then living and now famous artists.

The odd thing is that so many of them were born in the 1760s, or thereabouts, making it nearly as famous in the history of English art as the accession year, 1760, of George III was formerly believed to be in relation to the Industrial Revolution. Little can be more monotonous reading than a list of names, but a list of names there must be. Hogarth, Wilson, Stubbs, and Sawray Gilpin have already been noted as the early artists who were in a sense the forerunners of a movement that became more popular at a later date. George Romney (1734–1802) was in effect exclusively a portrait painter. There were others who were sporting painters, James Wootten (c. 1682–1764) amongst them. Men of the later age in this genre (and others) were Benjamin Marshall (1767–1835), James Ward (1769–1859) and J.F. Herring. George Morland depicted rural people in their usual physical surroundings, outside an inn, a cottage and so on (Pl. 17), but was not really painting landscapes, only rural circumstances. William Ward did much the same. "Old" Crome was born in 1768, and became the founder of the Norwich School, members being very numerous. But for landscape itself Constable is deservedly the best known if not the greatest master. There were many others, but they have been discussed very fully by the writers of art history, and omission from these pages does not mean that the artists omitted have no claim to fame, but it would be inappropriate to pretend to play the part of an art critic here, although something must be said about the subjects painted by these men. All of them chose to paint the rural scene, and it was the changes that were so rapidly being made in the appearance of the countryside that must have been their inspiration.

It is perhaps a trifle overweening to emphasise that the country as a whole was comparatively empty of population. All the people who conquered a large part of theworld in the eighteenth century did not number more than ten million, only the population of greater London today. This comparatively small number of people were mainly distributed in tiny villages of a few hundred inhabitants, rarely more. The picturesque and romantic scene was all about them as William Gilpin emphasised in his numerous works, and is shown in a slightly glorified setting by George Borrow at a rather later date

Oddly enough it is extremely difficult to find a picture showing open field cultivation, though some modern art experts write of this method of growing food as if it were the climax of human ingenuity. Unfortunately this halcyon period of English life was not generously illustrated in contemporary art. There is one depiction of work being done in fields of this type outside Cambridge in Loggan, *Cantabrigia illustrata* (Pl. 18). Most of the pictures that record its beauties belong to the end of the one and the beginning of the next century, and most depict enclosed fields, if they are not of pieces of waste land or of the solemn and forbidding mountain scenery beloved of those who adhered to the picturesque tradition that continued well into the nineteenth century. Grundy's words are worth quoting. "Efficiency was a mere cloud upon the horizon . . . the rural population was entirely static. There is no progress in a true golden age, and until the whole charming idyll dissolved in the tripartite revolution . . . of religion, farming and industry . . . a contented repose everywhere prevailed". [18] This optimistic statement could be expanded by more ample quotation. It is supported by the well-fed, happy looking people who figure in Morland's pictures, and in those of some others of his contemporaries; but there is a lot of evidence, written not pictorial, which denies it, and modern research has shown that the peasants and farmers who adorn these and other artists' pictures were in fact victims of a harsh and unrepentant system, kept up with rigour by the class that feared the effects of the French Revolution upon the body politic.

Besides the prosperity of the landowning classes resulting from their remodelling of the landscape and technical innovations, there were other incentives to production by the increasing number of artists as there were increasing numbers of all other classes. Amongst these were the well-to-do bourgeoisie of East Anglia who had always been interested in and buyers of the paintings produced in the Low Countries. They were the patrons of the Norwich School, which got well into its stride by the end of the century.

This school produced pictures of the improved landscape, some showing clearly the division of the land by hedgerows, and thus underlining the changes that had taken placc, but several of them were intrigued by the wild and desolate appearance of Mousehold Heath (Pl. 19), and some still adhered to the preposterously formidable scenes of the picturesque genre. East Anglian farming was particularly prosperous as well, landscape painters exceptionally numerous and perhaps exceptionally written about. This was truly what Kenneth Clark has denigrated. "The landscape of fact", he wrote, "like all portraiture is a bourgeois form of art",[19] a statement with which, if they were not dead, the ancestors of the nobility whose portraits dignify their own homes and the National Portrait Gallery, would not be very likely to agree. Landscape, for Clark, seems to have consisted in recognisable, unidealised views of their (the painters') own country, a sort of photograph in fact. Possibly these views were only recognisable by the spectator after the painters had pointed the moral. This sentiment possibly derives from Fuseli, who thought landscape entirely occupied with the tame delineation of a given spot.

If this was so it was nevertheless very appealing to a set of people whose taste had been trained by their overseas connexions, but they were not the only body among the bourgeoisie whose tastes for home decoration were being trained in the appreciation of art. These, in London, and possibly in the growing industrial cities, were not only class conscious, but were warmed by a spirit of emulation of people richer than themselves. These men formed a body of people who could be exploited, and where there are hencoops there is bound to be a fox, or several foxes. So it was in late eighteenth century London. Here there was a numerous and increasing class of people whose ambition may have been to possess original paintings but who were not financially well enough equipped to buy them: but they formed an expanding market for engravings and aquatints. They could not fail to be exploited any more than the helots of the artistic profession could. The number of ambitious young men who wished to demonstrate their powers, and thus to earn a living, was increasing, and the stimulant was there. J. Raphael Smith, son of a painter as he was, possessed power, and organised what can have been little more than a workshop for the production of engravings. He encouraged, exploited and made use of such men as William and James Ward and George Morland, and almost certainly others;[20] but while working for him they did get some kind of training, and when they broke away were able to earn a living by the sale of their own productions.

John Raphael Smith may broadly be described as a man who was the centre of a circle, to some members of which he was a mentor, and to others a parasite as well, but he was certainly a man of influence and a boon companion. He pretended to rate Morland low, but made an honest or dishonest penny out of this as of other connexions. He is also said to have obtained money from Rowlandson. His own pictures delineated very similar scenes to those of Morland.

The rural scene, whether including persons and things like carts and waggons and farm workers at work or taking their ease at their inn or home was not the only occupation that offered opportunity to the aspiring artist. Cattle breeding was being largely followed, and was becoming more and more profitable. Stubb's picture of the famous Lincolnshire ox has been noticed. Breeders of other animals wanted them put on record, and this created a market for the artist. Many did not miss this opportunity. Gainsborough was one; Sawrey Gilpin another; Sartorius, Reinagle, Morland and Rowlandson yet others; but perhaps the most important were James Ward and Benjamin Marshall. There were others who flourished in the days of the Regency and later.

Ward was caught in the net of the egregious Boydell (1719–1804), not perhaps by planned chicane, but by the kind of chance that happens in everybody's life. The Board of Agriculture, composed mainly of landowners, followed the pattern. It wished to have paintings of every breed of cattle, sheep and pig, done to measure, so as to make a permanent record of the development of the better class of animal currently being produced. They let a contract to Boydell to have the pictures made and engraved. Boydell employed James Ward to do the work for a set fee for each picture plus the necessary travelling expenses, but he failed to pay up. Ward made many pictures, but Boydell went bankrupt, and the project does not seem to have been completed. [21] Something of the same sort was undertaken as a private enterprise by George Garrard, and the results published in 1800. The attempt of the Board of Agriculture to have pictures made having failed, Garrard tried to supply the need and offered to make accurately measured scale models of the animals because he thought that a picture "rather exhibits a section or contour of the Animal than its real Image". [22]These models were made and exhibited, but the only surviving examples are to be found at Woburn Abbey and at Southill Park, near Biggleswade – so far as it has proved possible to discover. Garrard's book contains the portraits of several different breeds, and supplies a good idea of what the animals were really like.

Ten years before, Thomas Bewick had published his *General History of Quadrupeds*, 1790, and this led to his being asked to paint the portraits of some large and outstanding beasts, but since the owners were insistent upon having the really monstrous size of their favourites exaggerated in the portraits, Bewick, that strong minded man, refused this remunerative class of work. The Prince Regent commissioned Ben Marshall to make pictures of some of the cattle on the royal farms at Windsor in 1796.

One characteristic of the animal painters is a reflection of the times they lived in, and the mental attitude that their social circumstances created. Most of them, not content with the animals peacefully grazing or being looked at, or shown by their prideful owners, seem to have been overwhelmed by a need to paint pictures of animals fighting one another. Stubbs and James Ward are pungent examples. The combatants in these pictures are of the most unlikely. Of course lions must often have attacked horses (Pl. 20), perhaps boa constrictors too. The rhinocerous was too powerful and picturesque to be neglected. This preoccupation with fighting was one of the most potent in the human experience of the whole period. It was little wonder that the animal painters went in for the same sort of thing.

But there are always the optimists who can only see the past through their own rosy spectacles, and it is true enough that the aristocratic ruffians posessed redeeming features. "In 1803", says one rather optimistic writer, "England was still in the golden age; professional men like architects worked for educated patrons whose impeccable judgement of design was guided by their easy familiarity with the rules and ornamental conventions of the classic orders: the 'Gothic' taste was still an antiquarian diversion favoured by a few fastidious men of fashion . . . the English countryside was still almost untouched by organised industry". It was more beautiful and varied than any other, according to this patriot. From this year on, he writes, the nobility and gentry were untroubled by the European wars, pursued their normal life of pleasure, indulged their agreeable whims and eccentricities, built and furnished new houses, improved estates by planting and laid out great gardens. But there is always a dissident voice. Arthur M. Hind did not think so highly of eighteenth century taste. "It was a sophisticated age and unfertile soil for an art which needs above all things natural and forceful expression. It was an age of critics, connoisseurs and amateurs". [23] And again, "Most of William Woollett's work consists of landscape after Claude and his English imitators, Wilson, George and John Smith of Chicester. With William Byrne and other engravers working largely after Thomas Hearne, landscape developed into something little better than topography and need not concern us until we approach the school of Turner". This is an opinion it is impossible to agree with.

It is a temptation to say the truth is . . .; but when all is said and done the truth about any phase of history is what is thought about it by people looking backwards into time past and gone. All the elements making up the opinion of a certain time are complicated and inextricably intermixed. The great landowners desired at first pictures of their new houses and landscape gardens, something they could have put on record and be proud of. This led to an appreciation of the new landscape they were making, and that for itself. Not only did they wish for pictures of the immediate environment of their dwellings, but they wanted the changing countryside, as well as the prize animals they bred, recorded, and it may have been this that stimulated the English landscape painters of the late eighteenth century whose work has left so many delightful pictures, and who were indeed very numerous.

NOTES

1. R. H. Mottram, *Old England illustrated by English paintings of the 18th and early 19th centuries*, Studio, 1937.
2. See my *Old English Farming Books*, Vol. III, Pindar Press, 1983, pp. 181–237 and 260–265 for a description of the County Reports.
3. B. Taylor, *Animal painting in England from Barlow to Landseer*, Pelican, n.d.
4. On Marshall see my *Old English Farming Books*, Vol. II, 1950, pp. 114–120 and 152–153.
5. On these developments see my *The Farmer's Tools, 1500–1900*, 1952, pp. 170f.
6. C. S. Smith, "James Sharp: pioneer of rolling carts", *Country Life*, Dec. 7, 1972.
7. See W. Albert, *The turnpike road system of England, 1663–1840*, Cambridge University Press, 1972; Norman E. Lee, *Travel and transport through the ages*, Cambridge University Press, 1955; T.W. Wilkinson, *From track to by-pass. A history of the English road*, Methuen, 1934.
8. See, most recently, the *Illustrated Journeys of Celia Fiennes, 1685–1703*, ed. Christopher Morris, Macdonald, 1982.
9. E. Moir, *The discovery of Britain. The English tourists, 1540–1840*, Routledge, 1964.
10. A. Young, *Travels, during the Years 1787, 1788, and 1789*, Bury St. Edmunds, 1792, s.v.
11. See note 4 above.
12. R. Whitlock, "Setting the White Horse to rights", *Country Gentleman's Magazine*, June, 1972.
13. *Ibid*.
14. *Ibid*.
15. *Ibid*.
16. C. Hussey, *The Picturesque. Studies of a point of view*, Putnam's Sons, New York, 1927.
17. A. Young, *The poet and the landscape*, Hart-Davis, 1962.
18. C. R. Grundy, *English art in the 18th century*, Studio, 1928.
19. K. Clark, *Landscape into art*, Murray, 1949.
20. J. Frankau, *An 18th century artist and engraver. John Raphael Smith: his life and works*, Macmillan 1902.
21. See my *James Ward, R. A. Animal painter, 1769–1859, and his England*, Michael Joseph, 1974.
22. See my *Old English Farming Books*, Vol. III, Pindar Press, 1983, p. 35.
23. A. M. Hind, *A history of engraving and etching from the 15th century to the year 1914*, Constable, 1922.

IV

Post-War to the Repeal of the Corn Laws: the Florescence of Landscape Painting, 1815–1850

To add one more to the innumerable publications, both contemporary and recent, which discuss the state of the nation after Waterloo until the repeal of the Corn Laws, seems at first glance to be what might be called a work of supererogation: but in spite of the turmoil of these times, the appalling suffering of the industrial and agricultural slaves, the widening rift between the agricultural and industrial undertakers and workers, which was not then closed and indeed has never been since then, there was a continuation not only of agricultural enterprise and landowners' wealth, as well, it must be confessed, as a great deal of difficulty and distress amongst tenant farmers and small-scale landowners. The whole situation from 1815 to 1848 is indeed quite anomalous to the later observer.

So many books have been written about this period and so many learned men have spent themselves in describing it, either in part or in attempting a comprehensive record that it is impossible to read them all: but the point here is the connexion between the actions of the farmers and landowners in modifying the scenery of the countryside, the enormous effort that must have been made in creating the canals, and later the railways which superseded the canals, the ravages of industry in the north, in the Black Country, in South Wales and other places, so that apart from the continuing rural scene, very great changes were made in the appearance of the countryside with a consequent divergence between those artists who were enthused by the new industrial appearances, and those who remained devoted to what was still the preponderant mirror of England, the rural landscape as nature and man working together had made it.

Economic conditions are, of course, a determining factor in man's activities, and economic conditions pressed heavily on what has been spoken of as the agricultural interest for some thirty years following the great victory of Waterloo, a victory that was wildly celebrated by the mob, a circumstance that is said to have caused the Duke of Wellington to remark, "Now they are ringing their bells: soon they will be wringing their hands", a prophecy that proved only too true.

The task here is therefore to describe England and the English and how they lived between the military victory of Waterloo, and the comparatively peaceful social victories that were won between that day and the day of the repeal of the Corn Laws. It is not necessary to proceed beyond that date for the present purpose. The social background of the people who lived during this period has been sketched so many times and from so many varying angles that it would appear to be unnecessary to do this once more here: but, and the but is important, unless this is done it would be impossible to indicate the reasons why in a period of great social turmoil, such as men and women always resent, and a time when the two nations of Disraeli's novel became more widely separated than ever before, there was a great efflorescence of English landscape painting.

Widely as the life potential of the lord and servile tenant differed in the Middle Ages, they were, generally speaking, well known to each other. This continued to be true of the inhabitants of the villages of nineteenth-century England, their employees, the tenant farmers or landowners who farmed their own land, and overshadowing them all, the great estate. The tripartite social division was not invariable in every part of the country. There were variations, more especially in the north, the west, including Wales, and the southwest. It was only the very wealthy, the men who built the great houses, adorned them with landscape gardens and a variety of oddities, who could afford to buy pictures, and who had the developed taste to appreciate the talents of the growing number of painters, engravers and so on, who were prepared to work for them.

Immediately following the defeat of Napoleon there was little cause for rejoicing, as Wellington had underlined. The artificial prosperity of the landed interest during the twenty odd years of continuous warfare came to an abrupt halt with the peace. The workers had already been pauperised by the Speenhamland system and their situation could hardly have been made worse by the cessation of those hostilities that had created an artificial market, not only of the products of the farm, but for ships, cannon, shot and shell, gunpowder, and last but not least, the uniforms of the often unwilling soldiers and sailors who fought so well under penalties like flogging, and in spite of terrible conditions at sea with miserable and often decayed rations, owing to the peccadiloes of the contractors, and the corruption of the purchasing officers. Even so, these men, despite the immanence of violent death or disabling wounds, were probably better off than their peaceful brethren who stayed at home to produce the food and other goods.

Here it is the food producer that we are concerned with. The labourer's lot was distressful: his masters were to learn what it meant to produce for a falling market. With minor fluctuations this falling market continued for some twenty years, and the wonder is that any of the smaller estates survived as did some tenant farmers on the great estates. During this time there were several Parliamentary enquiries into the condition of the agricultural interest, and perhaps naturally more emphasis was laid upon the difficulties of the time than upon the successful methods adopted by some men in some places to overcome them.

The odd thing is that in a period of difficulty like this some men should turn their attention to the invention of novelties, or maybe it is natural rather than odd that men with acute minds should make the attempt instead of giving up to despair. Smith of Deanston designed his subsoil plough and combined it with a system of "thorough" drainage for the reclamation of wet lands. [1] A mole plough had been produced during the war for field drainage. The seed drill was gradually coming into more general use. Bell invented the first really viable reaping machine that was used in this country. The threshing machine was adopted on many farms, and became a favourite object of anathema in the "Captain Swing" riots. Heathcote designed a method of steam-driven ploughing that might have been a success but which failed mainly because of its massive structure and great weight. [2]

But despite the miseries and despair of the rural people it was in the new industrial towns that revolt against their pitiable state in those towns would soon break out. There were many agitators like the "Orator" Hunt, who planned to address a mass meeting in 1819, when the nervous yeomanry charged the defenceless people, and killed and wounded a great many while "dispersing the mob". [3] The abortive and possibly inspired Cato Street Conspiracy took place in the next year. The development and repression of the nascent trade union movement has been so often discussed that details are unnecessary here. The revolt of the farm labourers and the activities of "Captain Swing" are likewise too well known to need more than a mention. Its remorseless suppression was a measure of the fears of the ruling class as was the transportation of the Tolpuddle Martyrs a few years later. The Reform Bill, which gave the vote to some of the middle class, and the new Poor Law with its intention of ameliorating the appalling lot of the unemployed are other details of the history of the time that are to be garnered from any of the numerous textbooks and the specialist studies that have been devoted to these topics: but they had to be mentioned as they form the backdrop to the continuation, practically untouched so far, of the rule of the great landowners who were so powerful both in the areas they owned and indeed in the government of the country, although there were already cracks in this seemingly permanent facade of society.

It is unlikely that social organisation was ever really static. There must always have been dissident spirits who wanted change if only for their own advantage: their achievements were minimal perhaps, but they were able to cause, if not to make, modifications in the social structure, maybe not immediately apparent in their own day, but slowly becoming effective. Armed might was once the fulcrum of social organisation: financial might is that of the past two centuries if no longer, and continues to be so, which is perhaps to state something that is self-evident.

In the early decades of the nineteenth century this financial power resided in the hands of the great landowners, who not only had their rents as landlords of farming land, but royalties or direct payments for the industrial resources that were to hand on the surface of their estates, such as brick earth, lime and chalk for fertiliser, coal, iron and other minerals under the surface, and the provision of transport facilities upon the waters of the canals that were constructed through their estates.

All these activities made changes in the appearance of rural England as the continuation of the enclosure movement did. A further agricultural change was the exploitation of the heavy, wet land made possible by the invention of the mole plough, and more emphatically by the invention of a machine for making tubular draining pipes which took the place of the not too efficient tiles, half-moon shaped and resting on a flat base, or similar devices that had been previously used. The sort of thing that was going on is clearly shown by the frontispiece to G. H. Andrews, *Modern Husbandry*, 1853.

It is not certain that the depression of the landed interest was the direct cause of two events that happened in 1821 and 1822, but there is some reason for thinking so, The first of these two misfortunes was that Coke of Holkham held the last of his so-called "Sheep shearings" in that year. This climax to many years of such activities was not reached because of lack of success. The last of these grand undertakings was successful, and as well attended as any had been. The public were admitted to the estate and entertained for three days. A fourth day was limited to the house guests who listened to papers on agricultural subjects and farming improvements. The previous days had been devoted to an inspection of the farm's prize cattle and boars, novel implements and sheep shearing in the barn, and finally the carcasses of the beasts that had been slaughtered were examined, and a distribution of prizes made. On each of these days a public dinner was provided by Coke. The cessation of this annual event was a great loss to the farming community. [4]

The second event occurred in 1822. This was the dissolution of the semi-official Board of Agriculture that had been operating since 1793. The activities of this body had been of great value in recording the current practice of farmers in different parts of the country, and in propaganda for the adoption of novel systems found in the course of its explorations through the work of the writers of its County Reports, and more particularly perhaps the intensive work of its secretary, Arthur Young, and its first President, Sir John Sinclair. The former had died in 1820 after many years of blindness; the latter did not die until 1835, only three years before the foundation of the English Agricultural Society in 1838, a body that became the Royal Agricultural Society of England in 1839. This society, like the Bath and West, and the Highland and Agricultural Societies founded much earlier, did much to live up to its motto, "Practice with science", and has exercised enormous influence on the progress of farming in this country. It is renowned throughout the world. Many nobles and members of the landed gentry contributed essays to the *Journal* of the Royal Agricultural Society and those of the other Societies.

All these efforts were made to educate that recalcitrant being, the humdrum farmer, but at this time there was little opportunity for a farmer who desired it to secure any formal schooling in his profession. Some rather inefficient steps had been taken to supply this deficiency, some official, some private. A Chair of Agriculture had been established at Edinburgh in 1790, the Scots being enthusiastically inclined to all kinds of learning. A Chair of Rural Economy was set up in Oxford in 1796; but the lectures given were most certainly sporadic, and probably had little influence on the improvement of agricultural practice. About the Edinburgh Chair it would be bold to make any criticism. A Scotsman made the first attempt to provide agricultural education as a personal undertaking. This commendable person was J. C. Loudon, a man of indefatigable industry in his attempts to promote the interests of the agricultural or rural community. He established a school for farmers, where they would be able to learn the most up-to-date methods, at Great Tew in Oxfordshire. He took up residence in 1808. It was not the ordinary farmer that Loudon aimed for, His school was for sons of landed proprietors, and other young men who could afford to be trained for their life work on the land, perhaps as stewards, bailiffs and agents working upon great estates, or for those farmers who occupied the larger holdings. Amongst his other innumerable and prolix publications was one issued in 1809 with the title *Utility of agricultural knowledge to the sons of landed proprietors of England*. The school at Great Tew is reputed to have made a profit of £15,000 by 1812, a quite gargantuan sum in those days, and then Loudon abandoned the enterprise. He did not abandon others, and was certainly one of the most prolific writers of his day and age. Agricultural education as such did not really begin until the Royal Agricultural College was set up, an undertaking about which Dickens was enthusiastic. Even so, this school too was intended for what were then called "the sons of gentlemen", the inveterate snobbishness of early Victorian times.[5]

There had been some talk about setting up an experimental farm near London between various agricultural personalities, but although at one time a site had been considered, this proposal came to nothing. This had been one of the projects of the Board of Agriculture besides being privately discussed, i.e. as a desirable thing to have. Two members of the Board offered the use of land they owned for this purpose. Lord Carrington, when he was acting as President, offered the use of land he owned at High Wycombe, Bucks., and Sir Christopher Geary offered land near Croydon. The Board therefore dropped the idea of acquiring some land of its own until 1802 when six excessively expensive acres were taken on from Salisbury's Botanical Garden, Brompton. This action was condemned by Arthur Young, and the project was soon abandoned. No really scientific experiments were made until Rothamsted was founded by Lawes, assisted by Gilbert.

Meanwhile developments had been taking place quite without the benefit of education of a prescribed kind, and without the benefit of political pressure from government or government inspired authorities like the Board of Agriculture. Although transport facilities were elementary if judged by modern standards, they had been sufficiently improved to enable producers to send their goods for comparatively long distances, and this allowed of specialisation in farming so that the agriculture of Cheshire, for example, could be devoted, as it had more or less always been, to the production of dairy produce, cheese and butter, and as an incidental to calves without which the other things would not have been possible. This is but one example of the specialisation that was gradually becoming usual in appropriate areas. Other dairying areas were Cheddar, West Dorset, and Suffolk. Dorset and Suffolk butter were famous; Cheddar and Dorset cheese as well as a Double Gloucester and the famous Stilton were much appreciated, but Suffolk cheese was deplorably hard, and only suited for sale to the Navy. The great breeding areas for cattle and sheep were the West Midlands, Wales, Somerset, Sussex, East Anglia and the North. Horses were bred almost everywhere. The Cleveland Bay and the Suffolk Punch are outstanding examples. Corn crops were grown in the extensive areas of mixed farming, and exported to the increasing numbers of people in London, and the new or newish industrial towns. For these, too, the potato, that crop despised by William Cobbett, was grown in quantity along the banks of the Dee for the Irish immigrants and others in Liverpool and Manchester. Its consumption was spreading fairly rapidly. Vegetables and fruit were produced on the outskirts of the new towns, and more especially in Kent, Worcester, Hereford and the south-west where cider was made, sometimes from refuse apples, and sent to London and other places. Hops were extensively grown in Kent and the West Midlands, but the list is long and enough has been said to mark the increasing specialisation where farmers appreciated the necessity for producing for the market those products for which their land and climate were most suitable.

The essence of the new farming, whenever it began, either in the seventeenth or mid-eighteenth century, was to do what the pundits had preached from classical times. This was to try to return to the soil those elements that had been extracted from it by the growing plant, The best way to do this was by way of animal metabolism, i.e. to return animal waste products to the soil. This had been done for centuries by the sheep fold on the open fields, but not so efficiently as was possible by closer folding on enclosed fields, to which could be added the mass of cattle dung mixed with the straw bedding on which the cows were kept in yards or sheds. Mineral manures like chalk and lime and marl had also been used. Phosphates were supplied to grassland by the addition of crushed bones, etc. But the first imported fertiliser was so-called Peruvian guano, the massed droppings of seabirds on the islands where they bred. To this was added Chilean nitrate, and later the first true artificial, Superphosphate, in the 1840s; but this is not the place to go into minute details of farming practice.

All this of course had a marked effect upon the appearance of the countryside, the landscape, particularly where open fields or waste had been enclosed or where the farmsteads had been removed from the village street and rebuilt, or arranged in the midst or on the fringe of the enclosed fields that made up the holding. The Napoleonic wars had generated a great deal of patriotism though the results of the final victory had not turned out to be so very glorious for a major part of the population.

The only material change in the classification of the people was the addition of some more or less new elements, those who were engaged in industry, either as workers or as undertakers. Many of the latter were getting rich, with some vicissitudes, and aspired to become members of the ruling class, the almost but not quite closed ranks of the great landowners. Until the end of the period here being discussed it was the latter as representative, if not controllers of the landed interest, who dominated society. Incidentally they were the people who bought the native landscapes painted by the increasing number of local artists.

For the rest there was little change in social organisation. The rural population remained stratified as it had always been, if allowance is made for changes in terminology. Little or no change was made, except perhaps in the fashion of clothing, and, for the poorer classes, this was not material. Their poor vestments could hardly be dignified by the name of clothing, though the rustics look well enough provided for in the possibly idealised figures in the pictures of George Morland, James Ward, John Raphael Smith, and various artists of the Norwich School. Who must be believed, the painters or the written records with their statements about the issue of blankets, shoes and so on to the "villagers", whomsoever they may include? [6]

The immediate superiors of the labourers and cottagers were the tenant farmers working large or lesser acreages, the owners of what would now be called small-holdings, some of whom were little better off than the labourers, though they may have managed to scrape up a trifle better living, not perhaps in housing and clothing, but in food grown for home consumption, dairy produce for the same purpose, pork and bacon from the housed pig, perhaps some wool and mutton from a sheep or two; but between the farmers and the labourers there was an intermediate class of the subordinates of education. These were the unbeneficed clergy, and some others who were beneficed, but were of no more consequence to the great lords who possessed them – in the sense of being able to appoint or dismiss them. Poverty-stricken and oppressed as these men were, not to speak of their families, they may possibly have contributed to the statesman's ideal of having at least one gentleman, of minor or no consequence sometimes, in every parish. They were indeed a convenient venue for the promulgation of government edicts in a society that was largely illiterate.[7]

These men were not so impressive in village society as the substantial tenant farmers, most of whom rented farms on the great estates of the ruling class of landowners. Of course the general structure of rural society remained much what it had been before the wars until the 1860s, and indeed in some districts until the First World War, as H. G. Wells pointed out in his novel *Tono Bungay*; but though the structure remained much the same, the personnel changed – rather remarkably in some places. Estates had changed hands, a hazard that was inevitable when standards of living required more ostentation than the resources could support, but indeed the depression after the wars was fertile in personal change.

It had seemed difficult to understand how, when farms were being given up, and farmers (at least some, if contemporary reports can be believed) were being reduced to the status of labourers, the landowners seem to have been able to maintain their conspicuous expenditure. But this problem, not altogether easy to solve, has to what may be less simple minds an obvious explanation. During the Napoleonic wars great fortunes had been made in finance and industry. The Barings bought themselves into landed society; second generation Arkwrights became gentlemen; Sir Robert Peel was descended from an industrialist, and so on. This was indeed a blood transfusion into the veins of country life. There is no numerical account or arithmetical measure of the transfer of landed estates from hand to hand although there are some estimates, but the issue is confused by the sale of an estate by an owner and his immediate, or possibly not so immediate, purchase of another. All that can be said is that, as always, there was some transfer of landed property from one hand to another at this time, as at all others since the Norman Conquest.

The landowners great and small were not exactly drones battening on the body politic. They performed very serious jobs of administration, and were the governing body of rural society as well as enjoying leisure which enabled them to lead a cultivated life accompanied with rural sport, hunting, fishing and shooting as it was succinctly phrased in the early years of the present century. The admired manners of these aristocrats remained fairly static. They were formal and refined, brutal, drunken and cruel. At one and the same time a man could be a connoisseur of cockfighting (probably owning the birds), a shot who killed practically tame pheasants and partridges in large numbers, a patron of prize fighting, a bibulous creature whose health did not suffer, except in the most exaggerated cases like that of "Mad" Mytton, if he spent his nights under the table, but his days in the saddle chasing the fox or coursing the hare, or in wilder country, deer. He could also be a connoisseur of landscape gardening, proud of his house and grounds, but still more of his genealogy, and in consequence a collector and appreciator of *objets d'art*, and of the paintings of native landscape that English, with a severe emphasis on the word English, artists were producing to meet the taste of this many-sided and complex Englishman.

The changes that the activities of the enclosure movement and the new farming methods, both stimulated if not actually promoted by the most exemplary of the landowners, had made in the appearance of much of rural England not only supplied a reason for the landlords to wish it to be recorded in pictures but also created a demand and market for the work of the artists who made the paintings. Another thing it achieved was to make more hazards for the hunting man to overcome. To-day the dry stone walls of the Cotswolds, the western and northern counties and Wales, and the wide ditches dug for drainage purposes make interesting and often dangerous obstacles as some of the sporting artists were quick to perceive and depict. The popular Midlands were divided into fields by whitethorn, blackthorn and alder hedges, which had to grow to maturity before they formed effective restraints on the movements of grazing animals, and so were protected by wooden rails on both sides, creating what the foxhunting gentry called an "oxer", none too easy to jump, but interesting to the artist.

The great houses were staffed by hordes of servants usually recruited from the village families who lived on the estate. Day to day conduct of such an establishment demanded the services of both indoor and outdoor workers. Rank and place were definite among them as with their masters. Butler and housekeeper reigned over the cooks and housemaids with imperious authority. The stables were manned by coachmen, grooms and lads. The kennels were controlled by the huntsmen. A blacksmith's shop was essential. There was a small army of gardeners, whose business it was to produce such things as the prestige pineapples of which presents were made to underline the standing of the giver. Besides all this there might be a sawmill to supply material for repairs to house and outbuildings and fencing as well as a builder's yard for maintenance work upon the estate. Even in the nineteenth century such an estate was largely self-supporting, though clearly luxury goods must be imported as must material and equipment beyond local resources.[8]

The general management of such an estate required a great deal of legal business, and this might very well be taken care of by a steward or a firm of lawyers in a neighbouring town, although clearly there would be work for a London agent, no doubt a reputed firm of solicitors who might have succeeded to the business as regularly as the heir to the estate did so. Other services and facilities, too, were at hand in the market town where such people lived and had their being: but the family income was derived, to state the obvious, from the rents paid by the farmers who worked the land and produced the supplies of bread, meat and milk, besides the raw materials for many uses, such as hides for leather, wool for cloth, horn and hoof for glue, an almost inexhaustible list of what might be described as natural raw materials. The village also supplied services required by the farmers. Blacksmiths and wheelwrights were essential to the maintenance and repair of equipment, the making of tools, with the help of the villlage carpenter. Domestic requirements, few as they were, must also be provided for the village families, the farmer and his family, the labourer and his, the latter doubtless being minuscule.

The market towns from which the surrounding countryside was supplied with those necessities it could not provide for itself were the homes of a variety of trades that processed raw materials, for example, the saddler, the brewer, not forgetting the village publican whose premises were the scene of many activities besides drinking, the tanner, the miller, the tailor. The last did not provide the smock frocks of the labourers whose family clothing was quite often the cast-offs of more well-to-do people. The village "snob" (shoemaker) was quite capable of making the heavy footwear that country work made necessary.

It would be possible to enlarge this outline, but enough has perhaps been said to describe the social set-up in the period 1815–1850. Unfortunately some repetition is unavoidable because there had been little change in rural society. Some changes there were, as always, but they were only scratches on the substantial social facade that lasted so long.

At the risk of being tedious it is necessary to elaborate here the condition of rural England before the Reform Bill and the antecedent rebellion of the country yokels of the predominantly agricultural counties. As Chambers and Mingay put it, the essential function of the land was to supply a body of landlords who, with the urban middle class, some of whom were enfranchised by the Reform Bill, made up the governing class. [9] A few hundred landlords with extensive but sometimes scattered estates of many thousands of acres were able to dominate Parliament and the processes of government. The more numerous but lesser gentry owned together about half the area of the country, but played a subsidiary role. A few more were members of Parliament, but many were

J. P. s and therefore had control of country life, perhaps more especially in respect of the harsh and repressive Game Laws that failed to control the daring poachers of the villages. These writers quote a speech made by Brougham in 1816 when he said that wastes had disappeared for miles and miles giving place to houses, farms and crops, but he cast doubt upon the accepted canon of the result being that two blades of grass grew where one had grown before. Possibly, he thought, five then grew where four had formerly been established.[10]

This was of advantage to both farmers and landowners, but was of no help to the labourers. Their plight since Speenhamland had been pitiful; modern writers have reduced it somewhat, but all that their efforts have done is to localise the poverty and suffering that led to the Captain Swing riots which spread across England from Suffolk to Dorset and its repression with hanging and transportations. [11] Indeed at one point the Duke of Wellington, who relied on his military experience, wished to concentrate troops at Winchester to shoot down their fellows; much as he had done at the enclosure of Otmoor, though no shooting took place there.

Conditions for the labourer, the hind, were better in the northern counties where the arrangements were different, perhaps in diary counties like Cheshire where, according to Stella Davies, if working and in good health (two substantial provisos) the labourers family could be self-supporting, especially when the women earned a trifle by weeding, gathering stones, potato planting, spreading dung and filling up drains. [12] The New Poor Law changed the circumstances of the pauper. Instead of out-relief, confinement to a workhouse (sexes apart) was the punishment for this unplanned and unavoidable crime. "The yeoman", wrote Turberville, "had disappeared before the Reform Bill of 1832", with a trifle of exaggeration. [13]

Many villages retained their ancient flavour until the end of the nineteenth century, being then little different from what they had been for a long time, except of course for a railway that may then have connected them with the outside world when it could be reached. A pattern was Yarlington in Somerset which had then (1890) "always been as self sufficing and self contained as any village of its size with a population of about 220 and an area of some 1,200 acres can well be. Besides corn farmers and dairy farmers and farm labourers it can boast its own inn-keeper, miller, blacksmith, carpenter, mason, thatcher, sawyer and road contractor . . ." [14]

The amusements of such people were traditional and as regular as clockwork, being indulged at the appropriate and accepted time of the year. Prominent amongst these was the annual fair when a flock of strangers invaded the village to supply it with goods and amusement – at a cost. "The yokels had nothing to guide them but animal instincts", wrote Stella Margetson, "and sometimes they were cunning enough not to be fleeced by the sharpers waiting to pounce upon them", like the thimble rigger described in George Borrow's *Lavengro*, and though they sometimes woke up in a ditch, doubtless found this piece of dissipation (not dissimilar to that of their betters) a source of satisfaction.[15]

Most country people, farmers and labourers, village craftsmen, even the "officials", stewards, country lawyers and so on, were quite immobile. Travelling was not for them unless circumstances were very exceptional and demanded it. The carrier's cart and the travelling waggon were the usual means of transport, travel by foot being even more commonplace, but the roads were being improved, and the mail coach had been designed as early as 1742 though it did not come into moderately general use until the 1780s. The roads became better, and by the 1820s the stage coach was well established. Its era soon passed as the railway network was built. The canals too, as already said, carried heavy goods, but their use faded out before the onslaught of the railway, though relics of the system can be found in a derelict state here and there in the remote countryside, and some are still in use perhaps not always for the purpose for which they were at first intended. During the age of the stage coach side journeys, off the main routes, could be made in a post chaise hired with postillions from a local undertaking.[16]

Only the well-to-do could afford to travel by stage coach. It was costly and must have been tedious, and the drivers were not always the most civil of men. They considered themselves an elite, and carried themselves with consequence, practising the most consummate insolence towards the less important passengers, not always the "outsiders". There is no sufficient evidence to classify the passengers in these vehicles, and no valid assumption can be made about them. As to the nobles and superior landowners who made their way to London for the "season", lasting from January to June, many of them owned their own equipages and travelled in enormous style with outriders and baggage waggons carrying servants and supplies.

It was these men and women who set the fashions in London just as they ruled in the country, and in Westminster. They set the standards of taste, as well as the standards of many brutal and degrading pastimes. Their tastes and pleasures were a most extraordinary mixture of the refined and the revolting. Their psychology would certainly have interested Freud: but it is their refined tastes that are of most consequence here.

Though they ruled the country and continued to do so, in spite of the slight reduction in their powers made by the infiltration of the mercantile class into their ranks in a minor way, much of their time and thought was spent in fashionable pursuits. This was in effect the valueless and exhausting business of the unproductive consumption of luxuries and time, provoked by idleness of an exotic conventionality that only the people concerned in the performance of these prescribed rites can appreciate: but there is something to be said for this way of life. Its more admirable direction is the appreciation of the arts of literature, music and painting, as well as those devoted to the decoration of the person.

The vast scale of gambling on cards, dice and so on in the fashionable clubs, like Brooks, Whites, Crockfords, etc., has been sufficiently often described that it is well known to all interested in the period, and needs no elaboration here, except perhaps to remark that anything served as the foundation for a wager, ranging from horse racing, cock fighting, prize fighting and indeed almost any trivial happening.[17]

There was an extraordinary ambivalence in these noblemen and gentry. Enthusiastic as they were for what were then regarded as athletic sports, heavy drinking, overeating, and the sensual pleasures (as they are called) these cultured men could and did welcome with enthusiasm the development of English landscape painting that had come to its apotheosis in the work of the successors of Stubbs and Gainsborough, who were Turner, more especially Constable, and the unfortunates who became the helots of J. Raphael Smith, Boydell and Akermann, while in the eastern counties the Norwich school flourished, and its numerous members turned out innumerable pictures of the countryside in its varied aspects. This was only one facet of the intellectual life of the time. Poetry and literature flourished. But all this was the preserve of the "nobs", at least the intellectual setting; opposed was the "mob", as the common people were called, who had no representation in any branch of government from the minutest sections of local government to the magisterial processes of the House of Lords and the House of Commons.

The "mob" had no political voice. All it could do to show its pleasure or displeasure was to riot, and upon occasion riots could be suppressed with something more than a firm hand. The rioters sometimes had their revenge, physical if not political. The Captain Swing revolt of the agricultural labourers and its suppression has already been mentioned. The riots of the mob and the destruction caused is a commonplace of the history textbooks, but one incident is not without significance. A reform meeting of the populace was rather stupidly attended by a number of Tory gentlemen on horseback. They were howled at, and putting spurs to their horses fled at a gallop, but were pursued and overtaken in Marylebone fields where they cried havoc and were granted mercy with the fickle humour of all crowds. This did the people no good, any more than the riots at Bristol or smashing the windows of Apsley House. Here it is cited only as one aspect of behaviour on the part of the unenfranchised and the unruly and unruled (except by forces greater than their own) who were probably the greater part of the populace, the brutal and debauched multitude who had no other way of expressing their politics. Peel's new police had as yet no real power to control such demonstrations.

The food requirements of the increasing population of London, minimal as most of them were, was one facet of the expansion of the agricultural revolution to the heavy clays, very productive, and now possible to make dry with the new methods of drainage. Similarly the manufacturing population of the northern counties, the industries of the west midlands, the growing number of people employed in mining in the areas favoured by the mineral deposits made in total a much larger demand for the products of the farms, and encouraged the spread of the agricultural revolution (so-called) still further across the countryside, changing its appearance in a way that appealed to the recording artist who had evolved from his predecessors, the topographical and descriptive painters employed by the great landowners.

During the wars and after, the number of English artists was increasing. The Norwich School was flourishing, and their productions were excellent examples of the appearance of the countryside as it had been modified by the activities of improvers, as well as records of the wild appearance of the land that was as yet unreclaimed. In the Metropolis the activities of John Raphael Smith and his employees, if such they might be called, were responsible for much of the labours of apprentices like William Ward. By William's influence James Ward was employed by Smith, and so learned his trade, and laid the foundation of his future repute. George Morland, too, was exploited by Smith, and like the work of other painters many of his pictures were engraved and sold to the new class created by commerce and the progress of the industrial revolution, who were anxious to emulate richer and more well-born men in decorating their homes with examples of contemporary art.[18]

Many, if not most, of these artists depicted the rural scene in its domestic aspects, painting pictures of cottagers and farm workers at their tasks or in their rare moments of leisure. The wealthy sportsmen, landowners and gentry continued to have portraits of their racehorses made. They did not neglect their cattle. A great number of this genre has survived, and perhaps the sporting pictures and their painters may be discussed very briefly first. For example, Henry Alken (1785–1850) has been named by Aubrey Noakes as the "artist of the Corinthian age". This family was, he wrote, three generations of painters, engravers, etc. The war was not thought of by the renowned sportsmen, Osbaldiston, Assheton, Smith and others. They continued the "twilight revels of a doomed era" as if there was no war, and indeed went on doing so in the depression after the wars which is not a little difficult to understand, as difficult as the purchase of estates for what were then fabulous sums of money by the "new rich". Henry Alken's career, wrote Noakes, falls into two contrasting halves. His rise was rapid and satisfactory. Then he fell off . . . and somehow lost public favour.[19]

Ben Marshall was the outstanding example of this class of work. He has been condemned as a man without training, and therefore lacking in the fundamentals of his craft. This imaginary handicap did not prevent him being recognised by owners and officials at Newmarket, and modern opinion gives him high praise. In this kind of painting he was followed by John Frederick Herring who painted every winner of the St. Leger for about three decades. Herring also made portraits of cattle. James Ward was still painting cattle in 1820.[20]

These pictures of horses, cattle, sheep and pigs were not in the restricted sense landscape painting, but they were analogous, demonstrating as they did the interest of the ruling class in the living creatures, the horses that ministered to their sport and transport, the fine farm animals that they bred, perhaps by deputy through the home farm bailiff or the estate steward, even by their favourite and fortunate tenants. But though there was a galaxy of great landscape painters in the first few decades of the nineteenth century, this luminous age was doomed to decline into the sentimentality of Landseer and the later nineteenth painters in art, and Tennyson in poetry.

It is an accepted canon that the provincial success always wishes to repeat his or her success in the Metropolis. Many of the successful artists of this age corroborate this belief. Often they were men who were born in places separated from each other by what were then great distances; but the magnetic attraction of the ruling city was not of iron but of gold. Not only the commercial pre-eminence of the city with its wealthy merchants, but the high expenditure of the notables who came up for the season, emphasised the number and prosperity of possible buyers there.

But not all the artists suffered this magnet to attract them. The Norwich School at this time preserved its integrity and stayed put in its own countryside. The city itself, separated from London by a hundred miles or more of indifferent roads, had always looked more towards Europe, especially Flanders and Holland, and countries further afield for its commerce and culture. In these respects its leading citizens were as cosmopolitan as the merchants and bankers of London. Norfolk, too, was a county where many great landowning families possessed estates. These conditions gave the city an individual character all its own, and it is certain that this environment stimulated the artists of the school led by Crome who may be regarded as its founder and paterfamilias.[21]

The earlier interest of the citizens of Norwich in painters such as Rubens and Rembrandt has been remarked in a previous chapter. This interest was the foundation of their pride in and support of their locally born artists. The physical environment had a great deal to do with this. Norfolk was the county, *par excellence*, of the agricultural revolution, if such it can be called, and this agricultural revolution must have made remarkable changes in the appearance of the countryside. Some of it, at least, was tamed where it had been wild. George Borrow makes this clear in the descriptions to be found in *Lavengro*, but the Cliffords probably exaggerated in their book on *Crome* (1958), when they described the countryside of East Anglia as wild and picturesque. In the aesthetic theories of the time it would have been much less wild and picturesque than Wales or the Lake District. It was certainly much more completely cultivated, and by the then most modern methods, so that the rural scene in many parts of the county must have been one of cultivation in enclosed fields rather than the savage desolation of such places as Mousehold Heath, a favourite topic for local artists, an area that would have justified this terminology, but which was, like other such areas, comparatively restricted by this time.[22]

It is not my intention here to enter into the more psychological areas of art appreciation, but something must be said about modern comments on arts and artists. Nor is it my intention to enter into the influence of one artist upon another, a subject that has already been far too extensively discussed. It would be quite impossible to agree with the Cliffords that by the end of the eighteenth century no distinct tradition of landscape painting had materialised in England. This was rather obviously due to the fact that it was in the process of being developed. No exception can be taken to their theory, which is also held by others, and can hardly be controverted, that fundamental to the foundation of landscape painting was the topographical tradition in which landscape painting developed as an extension of the map or plan, perhaps also because some landowners, the possible purchasers, or people who commissioned the paintings, wished for something more than the somewhat arid maps and surveys of their estates. It must be admitted that the topographical artists were, or may have been, realists, so far as their capacity allowed, and that they tried to show a place as it actually was (in appearance to them or possibly in the appearance that their technique permitted) but certainly not schematised, though that is doubtful. This development is impossible to disagree with.

In every art there is a line of inheritance. The young practitioner must study his elders and forbears in that art, and consider whether he can introduce some novelty or improvement: but it is not intended here to make a study of the derivation of one artist's work from another, or his own personal development. Something of the sort may be essential, and must be touched upon, but this is not a study or critique of art. It only aims at showing the connexion between one particular human activity with others, and the stimulus of these other employments, farming and country life, upon the work of the landscape artists.

Landscape, like most other art, has in recent times been given all sorts of esoteric qualities. As long ago as 1902, C. J. Holmes suggested that it might be described as the pictorial interpretation of man's communion with nature. Laurence Binyon, at the end of the First World War, thought that the arts spring from a deep impulse and desire to create an ideal world more adequate to our innate conceptions than anything we can achieve in actual existence; but for him the century 1750–1850 was sterile, a period when spontaneous creation gave place to intelligent construction.[23] The artists and the poets knew very well, usually too well, what they were doing. It was a period when all the arts alike were content to observe rules.

This is a simpler explanation than all the subtlety. It is sordid. A market had developed, and the artists were quick to take advantage of it. The process of development that has been hinted at in earlier pages was that those who made the Grand Tour brought home examples of Italian, and possibly Dutch art, then with the building of great mansions, surrounded by landscaped gardens of wide area, their owners wished to have pictures of their achievements, the so-called topographical drawings. Then the face of the rural scene was changed by the enclosure of the open fields and wastes. This new appearance was attractive, and made possible the growth of a novel character. The progenitors of this genre were Richard Wilson, Gainsborough, Stubbs and their disciples as has been said. Apologies are offered for this repetition, but the continuity of the process deserves emphasis.

Members of the Norwich School, Ladbroke, Crome, Cotman, Thirtle, Robert Dixon and the rest were particularly fortunate in this respect. Not only were there still areas of uncultivated land in the country, but as every textbook of economic history emphasises there was much land that had been reclaimed or modified in its appearance by the famous improvers. In a sense therefore they had exceptional opportunities, in addition to a well developed sense of local patriotism, which indeed still exists.

The eastern counties were the birthplace of many famous exponents of landscape. Gainsborough was born in Sudbury, Suffolk, and Constable in the adjacent county of Essex, though Iolo Williams has argued that de Wint's landscapes were most typical of the English scene, a judgment that it is difficult to agree with.[24] When he said that James Ward had sometimes seemed to him like an aristocratic Cobbett he was not so far astray. Constable was a different man. He was a local pundit. He was enamoured of the rural scene, and his paintings have recorded it as it was for all time. One of his predominant characteristics was his appreciation of the pictorial value of canals and rivers, of canals made for the purpose of transport, of the value of watermills as a subject, and the modifications in the countryside that these things produced: but he was immune to the picturesque. He painted Stonehenge, the water meadows near Salisbury, George III's Weymouth, and several pictures of the robust wilderness of Hampstead Heath (Pl.21), besides his better known productions of local interest. Binyon indeed praised him as a conscious innovator who fought a decisive battle against the patrons and connoisseurs of his day with, it may be added, a good measure of success.[25]

Turner has been acclaimed as the foremost painter of his day, but he was not particularly interested, to judge from his productions, in the rural scene as a setting for human activities. His paintings, to express an uneducated opinion, are markedly impressionistic, and supply no detail that can readily be grasped, but his picture of a train on a viaduct, *Rain, steam and speed* (1844: Pl. 22), is a presage of a new day. It has been remarked that the Hanoverian period began in the days of horses and ended in that of the railway engine.

Samuel Palmer was no doubt a lesser light, but his paintings were a mirror-like reflection of the countryside, more particularly in Sussex, but examples could be multiplied. The phase changed about 1850, perhaps because of the changing aura of the Victorian age. The era of mechanism had been born.

London was of course the attraction it had always been and remains, despite all efforts to distribute excessive population into areas which will become densely populated quite rapidly. Naturally it was not the general population of London that bought pictures. It was the transient population of aristocratic families, the ruling class whose day was passing, who bought the pictures, more especially during the time they spent in the Metropolis following the exacting and expensive pleasures of the London season.

The nobility and gentry in their great houses usually had a gallery, a most pleasant but draughty place supplied with coal burning grates, and it was of the foremost prestige to have a row of portraits of ancestors hung upon the walls of this gallery. Other paintings, either imported or the product of native skill, may have formed part of the collection, which was not of course kept in London but in such ancestral halls as Chatsworth, Blenheim and Woburn Abbey. Pictures were valued not only as portraits of ancestors and living descendants, but of the countryside, no more that of it that was owned by the family than that owned by other persons, something the numerous picturesque works amply demonstrate.

Prestige symbols are no novelty. They are always in demand, whether they be the mechanical domestic devices used by a large proportion of modern people, or the paintings, elaborate and beautiful furniture, hangings and tableware owned by the aristocracy as fittings in their mansions, and being acquired by the bourgeoisie who were rising to power on foundations of substantial wealth. But rapidly as the changes were being made, they were not so remarkable as sometimes has been said. The increasing wealth of the numerous successful business men made them in some sense competitors of the long established landed aristocracy. Many of them wanted to buy themselves into this class, and succeeded in doing so. There was consequently some competition for the possession of the original paintings made by the large number of artists of first rate or lesser talents. This was more especially true of the cheaper engravings that enterprising (if no worse) men like John Raphael Smith were having produced for the new and expanding market. The middle class was prosperous and more than pleased to spend its money on those goods that would support, if not establish social standing; but it is a little severe to say, as Holmes did, that the paintings and engravings that they wanted could be manufactured by a trained artisan. [26] This is a theme that might well be elaborated, except for the fact that this has been done so by Hauser, whose conclusions are rather too definite. After 1830, he pronounced, the characteristics of the nineteenth century were already recognisable. The bourgeoisie is in full possession and awareness of its power. The aristocracy has vanished from the scene of historical events and leads a purely private existence. The victory of the middle class is undoubted and undisputed, [27] a formidable conclusion that is a trifle too assured.

During the Regency and the reign of George IV the growing wealth of certain members of the rising middle class is indisputable, and some of these fortunate creatures were buying themselves into the ruling class, but these events were on a comparatively minimal scale then, and the power and prestige of the nobility and gentry, the landowners great and small, had as yet hardly been breathed upon, that is before the passing of the Reform Bill of 1832.

The contradictory forces at play in Regency England and the following years were paralleled by the duality of masculine moral conduct in Victorian England. Its results are depicted in some of the narrative pictures of the time, but that was still to come in the second and third decades of the century. In these decades the leaders of society were afflicted by what might to-day be called split personality. It was not really so, of course, for even now the rich, both aristocratic and football fans, devote themselves to sport, and some people create a developed taste as well, or in spite of their other activities. These dual attitudes were much more definite in this earlier period. The upper class world was one of pronounced opposites, which seem to have been absorbed by individuals with no great difficulty. "Elegant refinement", as Lehmann put it, "went side by side with beastly cruelty, hard riding with Augustan scholarship, elaborate polish with insensitive coarseness, and artificial manners with boisterous practical jokes".

This duality of behaviour did not prevent the aristocracy from being the veneer of society. How could it? They were the rulers, they controlled much of the national wealth and resources. They were the patrons of painters and pugilists alike. For these people the London season was something almost inconceivable to any person born within the last half century. though it remained in a desiccated and possibly not so dissipated form until the cataclysmic ourbreak of the First World War in August, 1914. A century before that event the wealthy arranged and went to the so-called "routs" in the houses of great ladies. The men gambled and drank at clubs like White's, spent their afternoon quizzing great ladies of their own class and the most fashionable ladies of pleasure in Hyde Park, and even more equivocal dames in the darkness of Ranelaugh and Vauxhall, attending the exclusive society of Almack's, but curiously enough these dissipated people controlled the country and directed its fortunes. Not only were these men in command of policy, they were also possessed of skills that would have earned them a substantial living should they ever have been put to it. The rural sportsman knew enough to have got him a job as a gamekeeper, and members of the Four-in-hand Club could easily (and sometimes did) drive a stage coach; but it is unlikely that the men who sparred, doubtless with a care for aristocratic noses, more or less gently with the great professionals they admired so much, could have become pugilists. All these men were prepared to fight at the drop of a hat, nobility and slum ruffian alike. The renowned Assheton Smith did not disdain to fight with a coal heaver in Leicester because he thought himself insulted. This battle went on up and down the streets of the town until the local constabulary separated these two heroes.

This admiration of brutality was not confined to the classes; it was shared by the masses, and, astonishingly enough, by writers of the day like William Hazlitt, and a trifle later by that highly Christian gentleman and spreader of the Gospel, George Borrow – so it seems that violence and cruelty were commonplace throughout society. As a contrast to the really vivid appreciation of art, pictorial or literary, may be cited Hazlitt and Cruikshank. The picture of the second "Going to the fight" is definitely comparable with Hazlitt's literary description. This is worth repeating, or a small part of it, as an illustration of the mental attitude of a refined nature of that time.

At the fight, which Hazlitt described *in extenso*, one man was knocked down. Hazlitt wrote, no doubt for the delectation of his readers, "I never saw anything more terrific than his aspect just before he fell. All traces of life, of natural expression were gone from him. His face was like a human skull, a death's head, spouting blood. The eyes were filled with blood, the mouth gaped blood . . . Yet he fought on after this for several rounds", a truly elevating spectacle for a man of culture, a student of poetry.[28] Such a society was indeed highly sophisticated, and betted some £20,000 – an enormous sum in those days – on the result.

Men who could take punishment of this degree would put the modern pugilist to shame, and they were in consequence highly commended. George Borrow, son of a minor military officer, and a man of the most pronounced Christian principles, had nothing but praise for the bruisers of old England. As indicated, the younger members of the aristocracy found amusement in sparring with famous pugilists at a price, and it is pretty certain that the pugilists did not do to these men any great damage when earning their solatium. Even George Borrow, that hardy observer, did not care for another diversion, the rat pit, where dogs were confined in a pit, and rats put in for them to kill or be killed by. It is unnecessary to labour the point of the existence of senseless brutality in the ordinary lives of these men, coupled with the highest appreciation of the beauties of rural England and the painters who were making a permanent and beautiful record of that countryside.

These aristocrats, with their duality of mind, were the favoured of that social organisation. Beneath their glory was a festering world, a weltering social cesspit, the rookeries of the poor, workers or wastrels and criminals. Crime was then a desperate man's resource against the intolerable conditions of his life. Who would not have been a criminal in such circumstances? Such starvelings had no real conception of the life that was led by the rich although no doubt they flocked to see their entrances and exits as do modern people to a society wedding.

Neither the proud nor the humble could avoid the conditions in which the population of London and more particularly of the growing industrial towns were forced to live. The streets were inevitably polluted by the droppings of the innumerable horses, not to speak of the cattle, sheep and pigs that were driven through the streets of the Metropolis, and every other town. Even for the playboy and political aristocrat, life was something that few modern people would be prepared to accept. The atmosphere was foul with the smoke of coal fires, and with other very much less acceptable poisons. Lice and even bugs infested the houses of rich and poor alike. Pests of myriad flies were bred by the festering garbage of the streets. Food was contaminated to a monstrous degree. Milk was sold in open containers, and had usually been diluted with the water from the iron cow, the tap in every diary. Bread was poisoned with alum, chalk and other whiteners. The constituents of beer were more than suspect; but of course all this pressed more heavily upon the poor in their crowded and filthy ghettoes than on the rich in their fine houses in the West End. The fortunate amongst the workers were the vast numbers of domestic servants in an accepted hierarchy who were employed in the great houses. There were as many of these as there were agricultural workers enumerated in the Census of 1851. It is little wonder that Cobbett attacked Wilberforce for his propaganda for the black slaves of overseas when the native people of our island kingdom were in little or no better case. This is one of Cobbett's more reasonable apothegms, certainly more reasonable than his attack upon the uselessness of the British Museum when the vote for its maintenance was being discussed in the House of Commons. Cobbett went so far as to say that he did not even know where this useless haunt of the idle and lazy was.

The city was riddled by disease, naturally enough as a result of its insanitary streets, water supply, polluted for the most part, and the overcrowded houses. Enteric fever was an everyday affliction, the result of sewage disposal and food contaminated by foul ingredients. The high mortality of infants was sometimes attributed to the universal heaps of horse manure which caused infantile diarrhoea, a fly-borne affliction. Small-pox seems to have been endemic even until the late 1850s or later, despite vaccination. Fever was of equal potence. It was probably typhus. All this is not surprising, nor is the outbreak of cholera of 1832. It has been said that the life expectation of a boy born in 1841 was 40; a girl might look for two more years of life.

Between the upper and the nether millstones of the landed aristocracy and the squalid semi-starving poor, the new industrial and commercial men were increasing their wealth by their industry and frugality and were the prototypes of the successful men of the Victorian era. Compared with the insouciance of the man of "ancient lineage", these people were simple. Their behaviour was the result of the commendable and simple manners of the Quakers and Methodists, practising a strict and sometimes devout morality, and using the Bible as their handbook of living. A certain sterility of outlook may have been the result, but this was modified by an urgent desire to rise in the social world. Victorian snobbery is and has been for long a byword. One form this snobbery took is of consequence to the present theme. It made them the market for the engravings of the pictures that only the very wealthy could afford to buy, and which usually found a resting place in the mansions of the countryside.

It is true enough that much of London was a festering sore of misery, ill health and poverty, but it was comparatively a small city, and the country was within walking distance. There were fields at Kilburn, Islington and Hampstead, the Heath being a favourite spot with Constable, who painted several pictures of scenes there (Pl. 21). Landscape with pictorial values was on London's doorstep. North of the Oxford Road in Bloomsbury great landowners or their employees had built the series of squares and streets from Mecklenburg to Bloomsbury Square, a wonderful piece of town planning, and a lesson to our own times.

Could these squares have been the inspiration of Prinny and John Nash when they planned Regent Street, with the fine Quadrant at one end, Langham Place, Portland Place and the landscaping of Regents Park with its magnificent terraces? Unlikely, but possible. The Prince Regent may have had other ideas besides the wish to create a beautiful thoroughfare. He may have wished to separate the stews of Soho at the east side from the residences of the well-to-do on the west.

This is not the place to discuss the Prince Regent's character in detail, but he was the leader of society and his ideas and behaviour must have been a pattern that the members of his circle sought to emulate. Besides his love of architecture it is reported that he was a great admirer of Jane Austen, and kept a set of her books in each of his houses. On the opposite side of his character, he was lecherous, excessive in eating and drinking, in gambling and so on, a very pattern of the ambivalence of the members of the society of which he was the leader; but he bought paintings.

The apparent simplicity of taste (tinged with magnificence) that guided George IV was apparently shared by many of the class to which he in fact belonged. The immediate popularity of Bloomfield's poems, and the way he was lionised, is one proof of this. Of course, his poems like those of Clare that achieved equal success at a slightly later date, deal with country life, and this was the life that a very large proportion of English people continued to live, certainly until the middle of the century. This was the well from which was drawn the army of domestic servants mentioned above. No doubt the new industries welcomed recruits from the rural people. It is well known that in the beginning it was the children who were brought to what Blake called dark satanic mills, but which modern research has placed in a less harsh category. Of course the people of the new towns did not fail to reproduce themselves. They too were human.

By about 1850 the railway system was spreading everywhere. The reign of the coachmen was being mutinied against by the iron track and the steam train. In some places great landlords used their still preponderant influence to divert railway lines away from their villages. By-passed by the new railway, and no longer on the route of the vanishing stage coach, these peaceful towns and villages without railway stations returned to the sleepy lethargy of the pre-coaching era. The road no longer provided a spectacle of activity, the inns were deserted, and the nostalgic writers of today affect to believe that in consequence the smaller farmers and peasantry lived a tranquil and innocent life, a halcyon idea, but rather wide of the mark. Often village life was far from tranquil. There was an almost complete lack of amusements, except drinking and fornicating, and a continual succession of exacting tasks that recurred every day. When water was drawn from a well, it was a manual and time-consuming piece of work almost everywhere superseded today, though, where used, a semi-rotary pump takes time and work to fill the tank. There were none of the labour-saving gadgets that make life so much easier for everyone in the current absence of domestic help. The life of the rural working class was dour and hard. Work demands did not cease with the return from the fields. Animals must be fed and cared for, and there were domestic tasks. Boot repairs, carpentry of a domestic kind, and many other such tasks provided ample employment for the evening – if there was daylight enough – for candles and paraffin were costly items. Living in this way is far from poetic as Crabbe pointed out in the eighteenth century, though Cowper and Goldsmith idealised it.

There are wide areas of Europe today where the peasant lives in this primitive way, and the consequences are far from the appearances depicted in the country scenes of Morland, J. Raphael Smith, and others, though some of Henry Ward's pictures are definitely realistic. In modern Europe a different portrait of the peasant would have to be painted today, as it was in nineteenth-century France. The man and his wife and children are overworked and underpaid, just like his ancestors for centuries past. His careworn visage is a sign of the poverty of his life. His manual dexterity is won at the expense of disfigured, strongly manacled hands. His whole bearing and attitudes confirm the dullness and soul-destroying monotony of a poverty-stricken existence. Of course it is possible that the peasant does not realise his situation vis-a-vis his contemporaries of the machine-aided age.

Both the great age of the agricultural revolution and that of landscape painting were at an end by about 1850. Farming by that time was in general as up to date as the model farms of the early stages of the process had been. Most of the artists who had painted the landscapes the farmers made, and the landscapes of a previous period to be found scattered here and there throughout the countryside as well as the wildness of some of the still picturesque landscapes, were dead. A few lived until the next decade or so. Old Crome had died in 1821, Constable in 1837. Turner survived until 1851. Ben Marshall and John James Chalon, the animal painters, lived until 1835 and 1854. James Ward lived, but unproductively for the final years, until 1859. Peter de Wint died in 1849, David Cox ten years later.

The sublunary period of Victorian life had set in with the death of most of the outstanding artists of a masculine culture. They were succeeded by an insipid school which it would be improper to discuss here. This was reflected in the mawkishness of Tennyson, which earned him £10,000 a year with a peerage, and the sentimental work of Landseer whose romantic stags in reproduction graced the walls of my childhood home. It would be excessive for me to condemn these romantic stags and those pictures depicting social problems in the tone of Hogarth, but without his vivid poignance. It seems to be self-evident that something was lacking in the painters who practised in the latter years of the nineteenth century. Certainly severe criticism has been levelled by Basil Taylor on painters like J. F. Herring, but when all is said and done not everyone who practises one of the arts is necessarily a genius. Taylor stated that "His farmyards have the required shagginess of thatch and decay, but they are cleaner and more inviting, the animals are neater, the coats of the horses burnished, the ducks' plumage iridescent".[29] All this was an illumination if not a realistic picture of rural life. "The country must offer a respectable contrast to the towns", as he wrote, but Roger Fry thought that art sank to the level of triviality in the first half of the nineteenth century.[30] The unfortunate

Herring is accused of being the foundation of the decay which was made worse by Landseer, something I can very well agree with.[31] It was not subscribed to by Ruskin in his lectures at Oxford on "The art of England" in 1883.[32] For the present purpose enough has been said to demonstrate the close connexion between these two human activities, the one that produced the necessities of life, the other some part of its adornment.

NOTES

1. G. E. Fussell, *The Farmer's Tools, 1500–1900*, Melrose, 1952, pp. 68–69.
2. *ibid.*, p. 76.
3. On Peterloo see E. P. Thompson, *The making of the English working class*, Gollancz, 1963, pp. 680–700.
4. See A. M. W. Stirling, *Coke of Norfolk and his friends*, Bodley Head, 1912.
5. J. Gloag, *Mr. Loudon's England. The life and work of John Claudius Loudon*, Oriel Press, Newcastle upon Tyne, 1970.
6. J. B. Poynter, *Society and pauperism. English ideas on poor relief, 1795–1834*, Routledge, 1969.
7. A. T. Hart, *The curate's lot. The story of the unbeneficed English clergy*, John Baker, 1970.
8. See Mark Girouard, *Life in the English Country House: A Social and Architectural History*, Yale University Press, 1978.
9. J. D. Chambers and G. Mingay, *The agricultural revolution, 1750–1880*, Schoken Books, New York, 1966.
10. *ibid.*, pp. 126–7.
11. On the uprising see E. J. Hobsbawm and G. Rudé, *Captain Swing*, Lawrence and Wishart, 1969.
12. Stella Davies, *The agricultural history of Cheshire, 1750–1850*, Chetham Soc., 3rd ser., 10, 1960.
13. A. S. Turberville, *English men and manners in the 18th century*, Clarendon Press, 1929.
14. T. E. Rogers, *Records of Yarlington, being the history of a country village*, 1890.
15. Stella Margetson, *Leisure and pleasure in the 19th century*, Cassell, 1969.
16. N. E. Lee, *Travel and transport through the ages*, C.U.P., 1955; T. W. Wilkinson, *From track to by-pass. A history of the English road*, Methuen, 1934.
17. Stella Margetson, *Leisure and pleasure in the 19th century*, Cassell, 1969; *idem*, *Regency London*, Cassell, 1971.
18. Julia Frankau, *An 18th century artist and engraver, John Raphael Smith, his life and works*, Macmillan, 1902.
19. Aubrey Noakes, *The world of Henry Alken*, Witherby, 1952.
20. Basil Taylor, *Animal painting in England from Barlow to Landseer*, Pelican.
21. H. E. A. Day, *East Anglian Painters*, Vol I, Eastbourne Fine Art, 1967.
22. D. and T. Clifford, *John Crome*, Faber and Faber, 1958.

23. C. J. Holmes, *Constable and his influence on landscape painting*, Constable, 1902; Lawrence Binyon, *English poetry in its relation to painting and other arts*, Proc. Brit. Acad., Vol. VIII, 1918.
24. Iolo Williams, *Early English watercolours . . . artists born not later than 1785*, Connoisseur, 1950.
25. Lawrence Binyon, *English poetry in its relation to painting and other arts*, Proc. Brit. Acad., Vol. VIII, 1918.
26. C. J. Holmes, *Constable and his influence on landscape painting*, Constable, 1902.
27. A. Hauser, *The social history of art*, vol. II, Routledge, 1951, p. 715.
28. Wm. Hazlitt, *English comic writers*, Everyman ed. (1st, 1819).
29. B. Taylor, *Animal painting in England from Barlow to Landseer*, Pelican.
30. R. Fry, *Georgian art, 1760–1820*, Burlington Magazine Monographs, III, 1929.
31. *ibid.*
32. John Ruskin, *The art of England. Lectures given at Oxford*, 1883.

Bibliography

It seemed unnecessary here to specify the large number of contemporary books on farming, architecture, gardening and so on, and the massive writings of people like Sir John Sinclair, William Marshall and Arthur Young (see my *Old English Farming Books. Vol. III: 1793–1839*, Pindar Press, 1983), nor to detail the well-known modern writers such as Lord Ernle, C. S. Orwin and Joan Thirsk; but those of immediate relevance have been given in detail.

1. Abbey, J. R., *Scenery of Great Britain and Ireland . . . 1770–1860*, Curwen Press, 1952
2. Abbey, J. R., *Life in England in acquatint and lithography, 1770–1860*, Curwen Press, 1953
3. Albert, William, *The turnpike road system of England, 1663–1840*, Cambridge University Press, 1972
4. Allom, T., and Pickering, G., *Lake and mountain scenery . . . Westmorland and Cumberland*, n. d., c. 1833
5. Armitage, Doris May, *The Taylors of Ongar*, Heffer, 1939
6. Ashton, William, *Social England under the Regency*, 2 vols., 1890
7. Ashton, William, *When William IV was king*, Chapman and Hall, 1896
8. Austen, Jane, *Memoir by her nephew James Edward Austin-Leigh*, 1926 (1967: first published 1870)
9. Allen, W. Gore, *Renaissance in the North*, Sheed and Ward, 1946
10. Ashton, T. S., *An economic history of England: the eighteenth century*, Methuen, 1955
11. Barley, M. W., *The English farmhouse and cottage*, Routledge, 1961
12. Barrell, John, *The idea of landscape and the sense of place, 1730–1840*, Cambridge University Press, 1972
13. Batchelor, Denzil, *The turf of old*, Witherby, 1951
14. Bayne-Powell, Rosamund, *Travellers in 18th-century England*, 1951
15. Bedford, Duke of, *A great agricultural estate*, Murray, 1897
16. Bewick, T, *General history of quadrupeds*, 1790
17. Binyon, Lawrence, *English poetry in its relation to painting and other arts*, Proc. Brit. Acad., Vol. VIII, 1918
18. Binyon, Lawrence, *The followers of William Blake*, Halton etc., 1925
19. Bovill, E. W., *The England of Surtees and Nimrod*, Oxford University Press, 1959
20. Bovill, E. W., *English country life, 1780–1830*, Oxford University Press, 1962

21. Bowness, Alan, "Art and society in England and France in the mid-19th century", *Trans. Roy. Hist. Soc.*, 5th ser., 22, 1972
22. Boxer, C. R., "The Dutch economic decline", in Cipolla, *The economic decline of Empires*, Methuen, 1970
23. Burton, Elizabeth, *The Georgians at home, 1714–1830*, Longmans, 1967
24. Bury, Adrian, *Two centuries of British water colour painting*, Newnes, 1950
25. Carpenter, Patrick and Graham, *Art and ideas. An approach to art appreciation*, Mills and Boon, 1971
26. Chambers, J. D., *Nottinghamshire in the 18th century*, 2nd. ed. 1966
27. Chambers, J. D., *Population, economy and society in pre-industrial England*, Oxford University Press, 1972
28. Chambers, J. D. and Mingay, G. E., *The Agricultural revolution, 1750–1880*, Schoken Books, New York, 1966
29. Christie, O. F., *The transition from aristocracy, 1832–1867*, Seeley Service, 1927
30. Clark, Kenneth, *Landscape into Art*, Murray, 1949
31. Clarke, Ernest, "Agriculture and the House of Russell", *Journal of the Royal Agricultural Society of England*, 1891
32. Clifford, Derek, *Water Colours of the Norwich School*, Cory Adams and Mackay, 1965
33. Clifford, Derek and Timothy, *John Crome*, Faber and Faber, 1958
34. Coleman, D. C., "Gentlemen and Players", *Economic History Review*, 2nd ser., XXVI. i., Feb. 1973
35. Cooper, T. S., *Thirty four subjects of cattle*, 1837
36. Cooper, T. S., *Drawing book of animals and rustic groups*, c. 1860
37. Cowie, L. W., *Hanoverian England, 1714–1837*, Bell Modern Histories, 1969
38. Cundall, H.M., *History of English water colour painting*, Murray, 1908
39. Darvall, Frank Ongley, *Popular disturbances and public order in Regency England*, Oxford University Press, 1934
40. Davies, Randall, *English society in the 18th century in contemporary art*, Seeley and Co., 1907
41. Davies, C. Stella, *The agricultural history of Cheshire, 1750–1850*, Chetham Soc., 3rd ser., 10, 1960
42. Day, Harold E. A., *East Anglian Painters. Vol. I,* Eastbourne Fine Art, 1967 (?)
43. Day, Harold E. A., *East Anglian Painters. Vol. II, Norwich School*, Eastbourne Fine Art, 1968
44. Day, Harold E. A., *East Anglian Painters. Vol. III*, Eastbourne Fine Art, 1969
45. Derry, John W., *Reaction and reform, 1793–1868. England in the early 19th century*, Blandford, 1963
46. Driver, Edmund, *Cheshire and its cheese makers, their homes, landlords and supporters*, Blandford, 1963
47. Edwards, Edward, *Anecdotes of painters who have resided or been born in England*, 1808
48. Falk, Bernard, *Thomas Rowlandson: his life and art*, Hutchinson, n.d.
49. Frankau, Julia, *Eighteenth century colour prints*, 1900
50. Frankau, Julia, *An 18th century artist and engraver. John Raphael Smith: his life and works*, Macmillan, 1902

51. Fry, Roger, et al., *Georgian art, 1760–1820*, Burlington Magazine Monographs, III, 1929
52. Garrard, George, *A description of the different varieties of oxen common in the British Isles . . .* , 1800
53. Gaunt, W., *English Rural Life in the 18th century*, Connoisseur, 1925
54. Gaunt, W., *The great century of English painting. Hogarth to Turner*, Phaidon, 1971
55. Gilbey, Sir Walter, *Animal painters of England from the year 1650*, Vinton and Co., 1900
56. Gilbey, Sir Walter, and Cuming, E. D., *George Morland, his life and works*, Black, 1907
57. Gloag, John, *Two thousand years of England*, 1952
58. Gloag, John, *Mr. Loudon's England. The life and work of John Claudius Loudon*, Oriel Press, Newcastle on Tyne, 1970
59. Gough, J. W., *The rise of the entrepreneur*, Batsford, 1969
60. Grant, Col. M. H., *A chronological list of the old English landscape painters (in oil) 16th–19th centuries*
61. Grossmann, F., *Bruegel. The paintings*. Phaidon, 1955
62. Grundy, C. Reginald, *English art in the 18th century*, Studio, 1928
63. Haraszti-Takaco, Marianne, *Rubens and his age*, Corvina Press, Budapest, 1972
64. Hardie, Martin, *Water colour painting in Britain. I. The 18th century*, Batsford, 2nd edition, 1969
65. Harris, Alan, *The rural landscape of E. R. Yorks, 1700–1830. A study in historical geography*, 1969, 1st edition, 1961
66. Harris, R. W., *Romanticism and the social order, 1780–1830*, Blandford Press, 1969
67. Hart, A. Tindal, *The curate's lot. The story of the unbeneficed English clergy*, John Baker, 1970
68. Hart, W. H., *Selections from the correspondence of Robert Bloomfield. The Suffolk poet*, 1870, reprinted 1968
69. Haslam, C. S., *The biography of Arthur Young* (to 1787), George Over, Rugby, 1930
70. Hauser, Arnold, *The social history of art*, vol. II, Routledge, 1951
71. Hazlitt, William, *English comic writers*, 1819
72. Hewlett, Esther, *Cottage comforts, with hints for promoting them gleaned from experience*, 1825
73. Hind, Arthur M., *A history of engraving and etching from the 15th century to the year 1914*, Constable, 1922
74. Holderness, B. A., "Landlord's capital formation in East Anglia, 1750–1870", *Economic History Review*, 2nd series, XXVIII, August 1972
75. Holmes, C. J., *Constable and his influence on landscape painting*, Constable, 1902
76. Howe, G. Melvyn, *Man, environment and disease in Britain. A medical geography of Britain through the ages*, David and Charles, 1972
77. Hughes, C. E., *Early English watercolour*, revised and edited by Jonathan Mayne, Benn, 1950 (1st edition 1913)
78. Hutchinson, Sir Joseph, *Farming and food supply. The interdependence of countryside and town*, Cambridge University Press, 1972

79. Hussey, Christopher, *The Picturesque. Studies of a point of view*. Putnam's Sons, New York, 1927
80. Hussey, Christopher, *British Country Life in Art*, Country Life, 1937
81. Hyams, Edward, *Capability Brown and Humphrey Repton*, Dent, 1971
82. Inglis, Brian, *Poverty and the Industrial Revolution*, Hodder and Stoughton, 1971
83. Klingeneer, Francis D., *Art and the Industrial Revolution*, Noel Carrington, London, 1947
84. Laslett, Peter, *The world we have lost*, Methuen, 1965
85. Lee, Norman E., *Travel and transport through the ages*, Cambridge University Press, 1955
86. Llewellyn, Alexander, *The decade of reform. The 1830's*, David and Charles, 1972
87. Lutyens, Mary, *The Ruskins and the Grays*, Murray, 1972
88. Malins, Edward, *English landscaping and literature, 1660–1840*, Oxford University Press, 1966
89. Manwaring, Elizabeth Wheeler, *Italian landscape in 18th century England . . . the influence of Claude Lorrain and Salvator Rosa on English taste, 1700–1800*, Cass, 1965 (1st edition, 1925)
90. Margetson, Stella, *Leisure and pleasure in the 19th century*, Cassell, 1969
91. Margetson, Stella, *Leisure and pleasure in the 18th century*, Cassell, 1970
92. Margetson, Stella, *Regency London*, Cassell, 1971
93. Marshall, Dorothy, *English people in the 18th century*, Longmans, 1956
94. Miller, Thomas, *Turner and Girtin's picturesque views sixty years since*, 1854
95. Mingay, G. E., *English landed society in the 18th century*, Routledge, 1963
96. Moir, Esther, *The discovery of Britain. The English tourists, 1540–1840*, Routledge, 1964
97. Morley, Henry, *First sketch of English literature*, 13th edition, 1887
98. Mottram, R. H., *Old England illustrated by English paintings of the 18th and early 19th centuries*, Studio, 1937
99. Morus (Richard Lewisohn), *Animals, men and myths. A history of the influence of animals on civilisation and culture*, Gollancz, 1954
100. Nimrod (Charles Apperley), *The life of a sportsman*, John Lehmann, 1948
101. Noakes, Aubrey, *The world of Henry Alken*, Witherby, 1952
102. Parker, Constance Anne, *Mr. Stubbs, the horse painter*, J.A. Allen and Co., 1971
103. Plumb, J. H., *The death of the past*, a Pelican book, c. 1972
104. Poynter, J. R., *Society and pauperism. English ideas on poor relief, 1795–1834*, Routledge, 1969
105. Prideaux, S. T., *Aquatint engraving. A chapter in the history of book illustration*, Duckworth, 1909
106. Reed, Donald, *The English provinces, c. 1760–1960, a study in influence*, Ed. Arnold, 1964
107. Redgrave, Richard and Samuel, *A century of painters of the English school*, Smith Elder, 3 vols., 1866
108. Reeve, Clara, *The old English baron*, 2nd. edition 1778, ed. with an introduction by James Trainer, Oxford University Press, 1967
109. Reitlinger, Gerald, *The economics of taste. The rise and fall of picture prices, 1760–1960*, Barrie and Rockcliffe, 1961

110. Rogers, T. E., *Records of Yarlington, being the history of a country village*, 1890
111. Roscoe, E. S., *The English scene in the 18th century*, Constable, 1912
112. Ruskin, John, *Lectures on landscape delivered at Oxford*, 1871
113. Ruskin, John, *The art of England. Lectures given at Oxford*, 1883
114. Russell, Ronald, *Lost canals of England and Wales*, David and Charles, 1971
115. Salaman, Malcolm C., *Old English colour prints*, London, 1909
116. Sparrow, W. Shaw (ed.), *In rustic England*, Hodder and Stoughton, 1906 (pictures by Birkett Foster, 1825–1899)
117. Sparrow, W. Shaw, *A book of English etchings from Francis Barlow to Francis Seymour*, John Lane, 1926 (156 plates)
118. Sheppard, Francis, *London 1808–1870. The infernal wen*, Secker and Warburg, 1971
119. Shoberl, Frederic, *Natural history of quadrupeds . . . illustrated by Thomas Landseer*, 2 vols., 1834
120. Siltzer, Capt. Frank, *The story of British sporting prints*, Hutchinson, n.d.
121. Smith, C. S., "James Sharp: pioneer of rolling carts", *Country Life*, Dec. 7, 1972
122. Steegman, John, *The artist and the country house*, Country Life
123. Stephen, Leslie, *English literature and society in the 18th century*, Duckworth, 1910 (Ford Lectures, 1903)
124. Stevenson, R. L., *Essays on travel. An autumn effect*, 1875
125. Stirling, A. M. W., *Coke of Norfolk and his friends*, Bodley Head, 1912
126. Stroud, Dorothy, *Henry Holland, 1745–1806*, Art and Tecnics, 1950
127. Stroud, Dorothy, *Capability Brown*, Country Life, 1950
128. Stuart, D. M., *Regency roundabout*, Macmillan, 1943
129. Taylor, Basil, *Animal painting in England from Barlow to Landseer*, A Pelican Book, n. d.
130. Taylor, Basil, *Stubbs*, Phaidon, 1971
131. Thomas, David, *Agriculture in Wales during the Napoleonic Wars*, University of Wales Press, 1963
132. Thompson, F. M. L., *English landed society in the 19th century*, Routledge, 1963
133. Turberville, A. S., *English men and manners in the 18th century*, Clarendon Press, 1929
134. Walker, Stella A., *Sporting art: England 1700–1900*, Studio Vista, 1972
135. Walpole, Horace, *Anecdotes of painting . . . incidental notes on other arts collected by George Vertue*, 2nd edition, 1765
136. Webber, Ronald, *The village blacksmith*, David and Charles, 1971
137. Whitlock, Ralph, "Setting the White Horse to rights", *Country Gentleman's Magazine*, June, 1972
138. Whitman, Alfred, *The masters of mezzotint. The men and their work*, Bell, 1898
139. Wickett, William, and Duval, Nicholas, *The Farmer's Boy. The story of a Suffolk poet. Robert Bloomfield*, Lavenham, 1971
140. Wilkinson, T. W., *The highways and byways of England*, Iliffe, 1913
141. Wilkinson, T. W., *From track to by-pass. A history of the English road*, Methuen, 1934
142. Williams, Iolo A., *Early English watercolours . . . artists born not later than 1785*, Connoisseur, 1950

143. Woodforde, John, *The truth about cottages*, Routledge, 1969
144. Young, Andrew, *The poet and the landscape*, Hart-Davis, 1962
145. Zegrosser, Carl, *A book of fine prints*, Crown Publications, New York, 1937

Index

ACKNOWLEDGEMENTS

Copyright of the following photographs is acknowledged: Iveagh Collection, Pl. 10; Kunsthistorisches Museum, Vienna, Pls. 2, 4; Lever Collection, Port Sunlight, Pl. 16; McFadden Collection, Philadelphia, Pl. 9; Metropolitan Museum of Art, New York, Pl. 1; National Gallery, London, Pls. 12, 13, 22; Norwich Castle Museum, Pl. 19; Royal Academy London, Pl. 21; Tate Gallery, London, Pls. 8, 14, 20; Walker Art Gallery, Liverpool, Pl. 11.

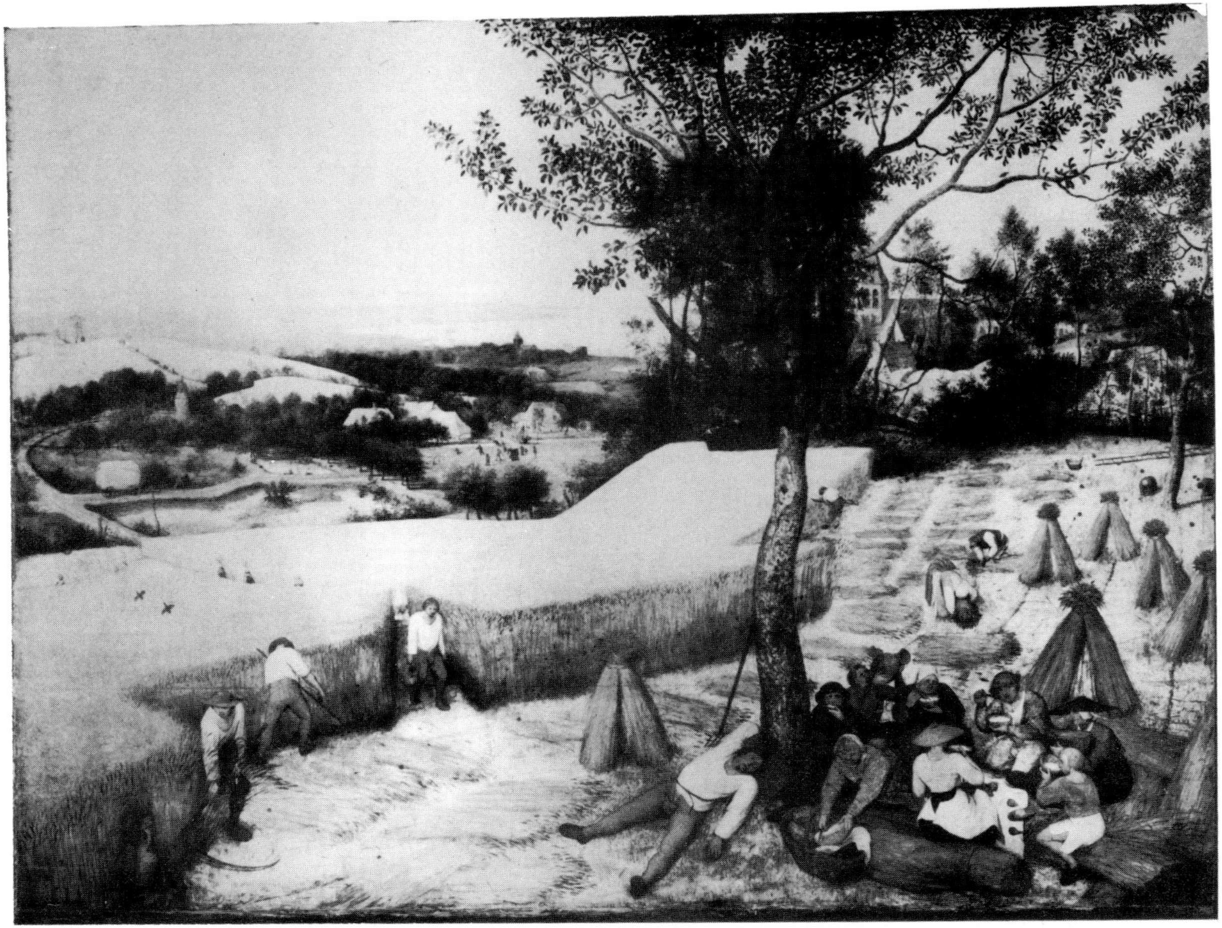

Pl. 1. Pieter Bruegel, *Harvesters*
Metropolitan Museum of Art,
New York

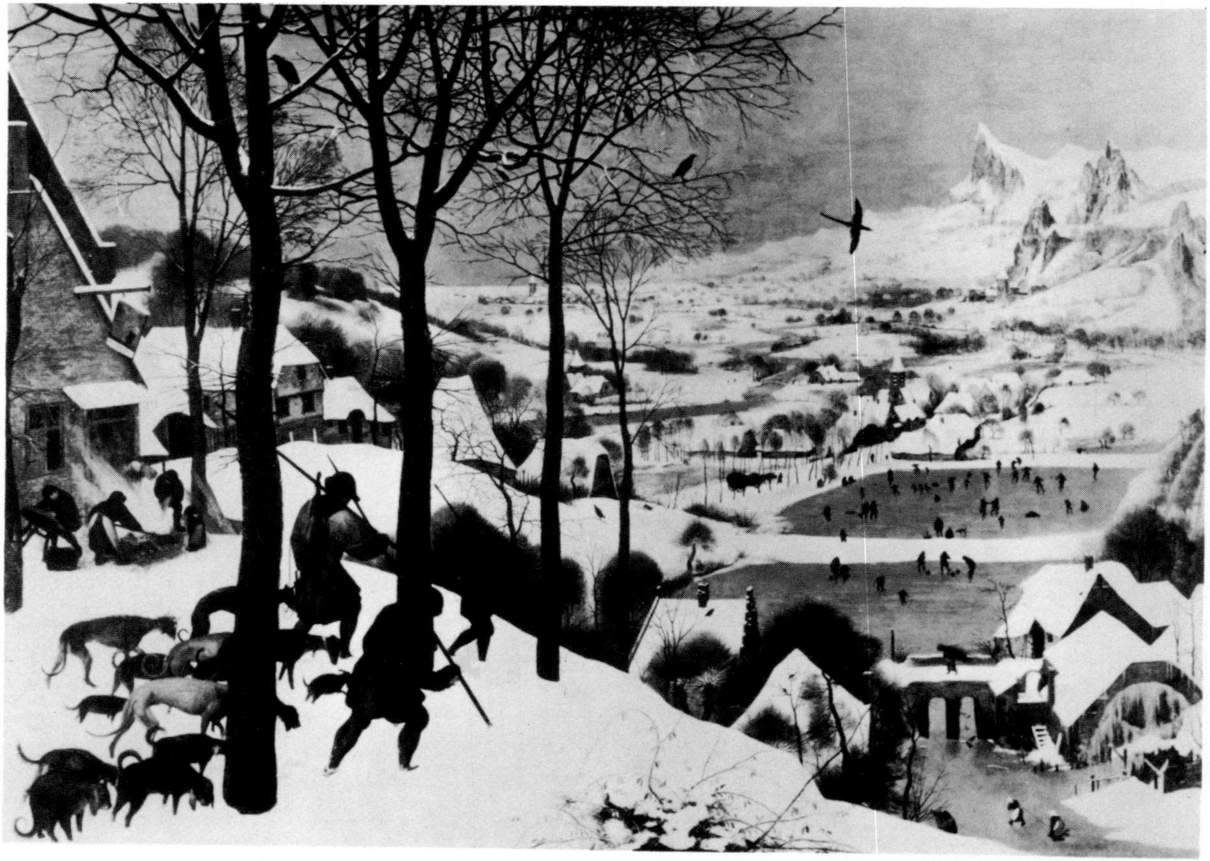

Pl. 2. Pieter Bruegel, *Winter*
Kunsthistorisches Museum, Vienna

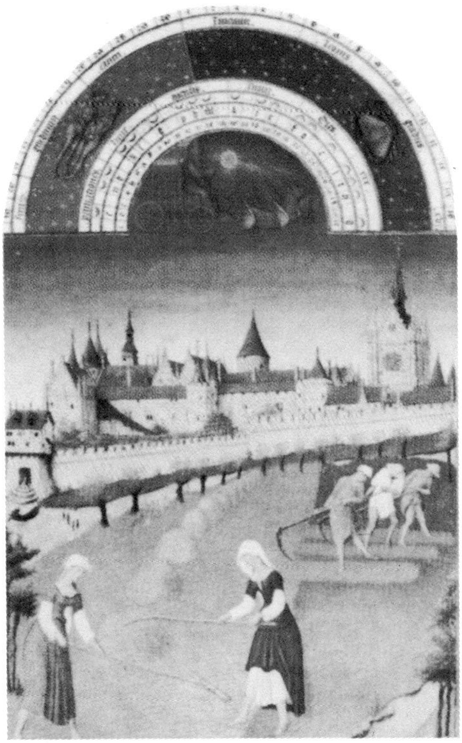

Pl. 3. Miniature from *Les Très Riches Heures du Duc de Berry*, Musée Condé, Chantilly

Pl. 4. Pieter Bruegel, *The Return of the Herd*, Kunsthistorisches Museum, Vienna

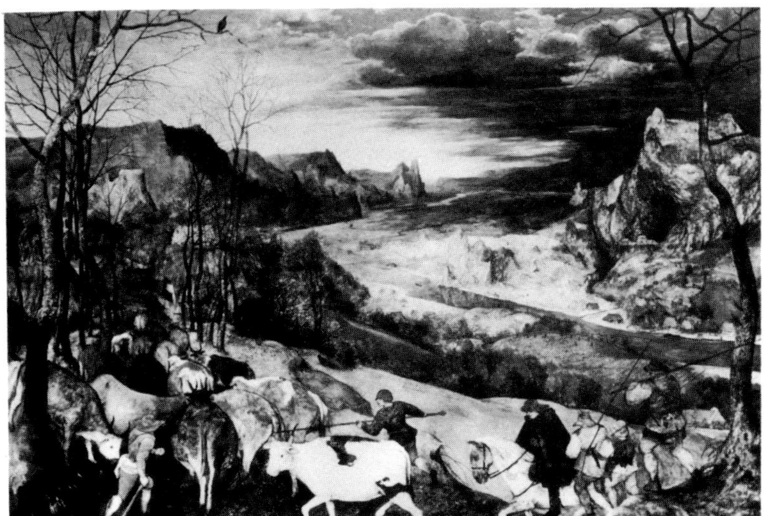

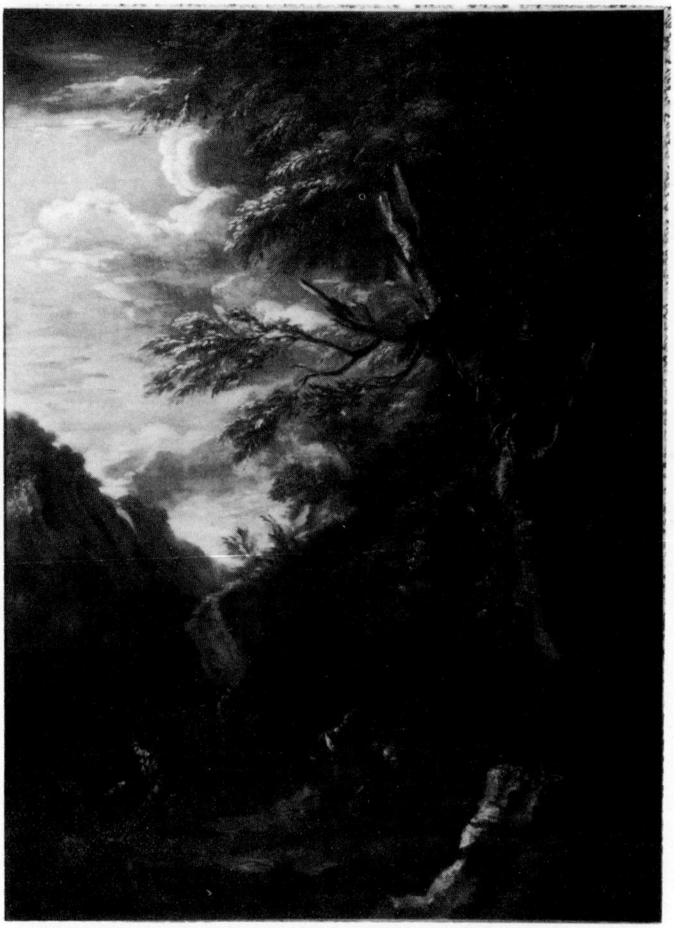

Pl. 5. Salvator Rosa, *Mountain Landscape*

Pl. 6. Francis Place, *The Dropping Well at Knaresborough*
British Museum

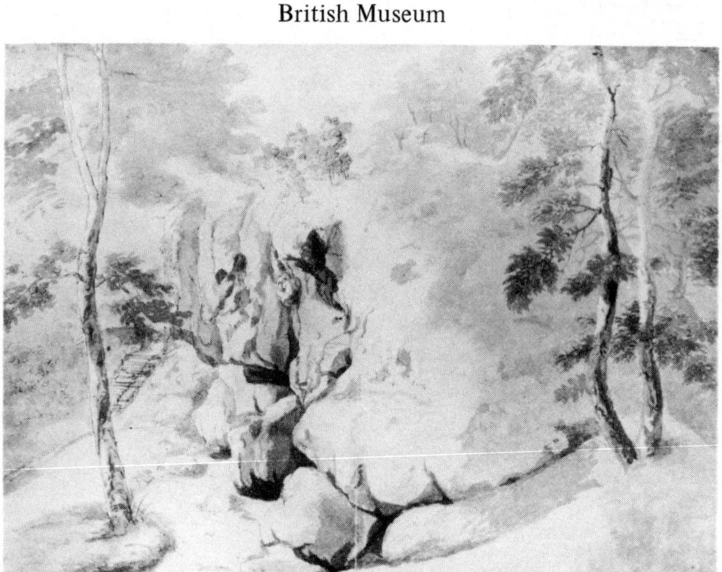

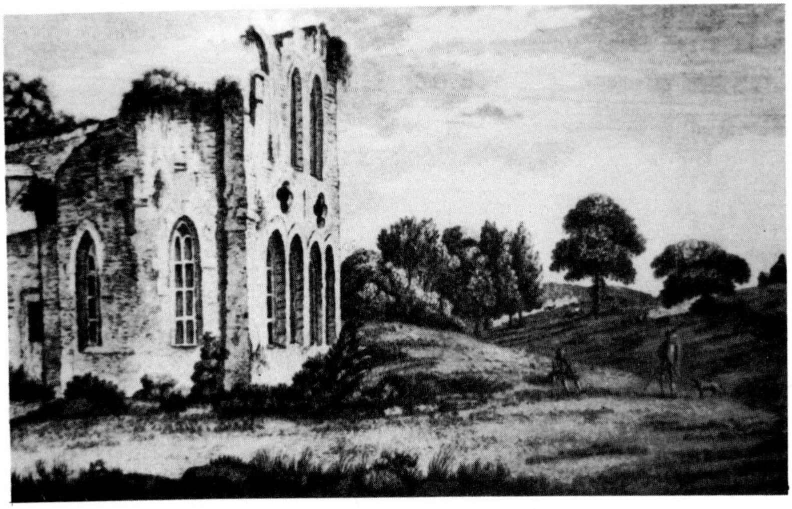

Pl. 7. *The Leasowes* engraved by D. Jenkins, 1779

Pl. 8. Richard Wilson, *Hounslow Heath*,
Tate Gallery

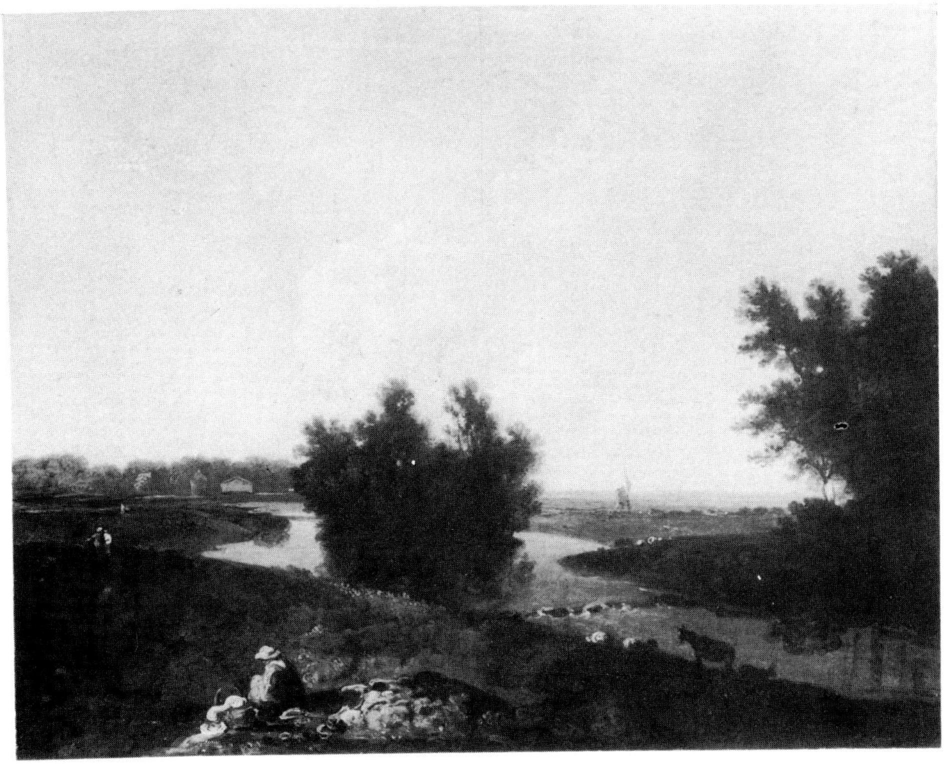

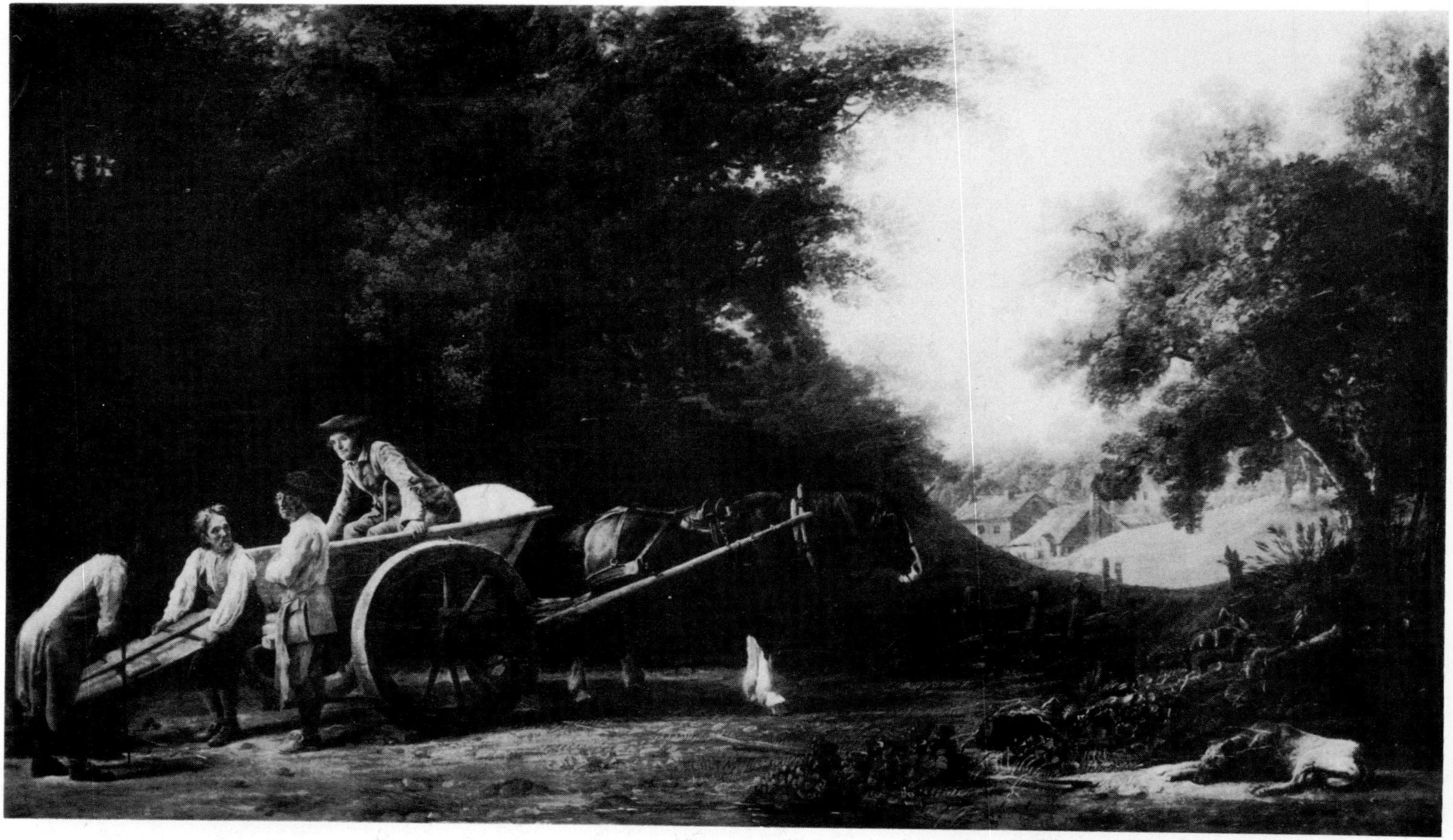

Pl. 9. George Stubbs, *The Brick Cart*
Mc Fadden Collection, Philadelphia

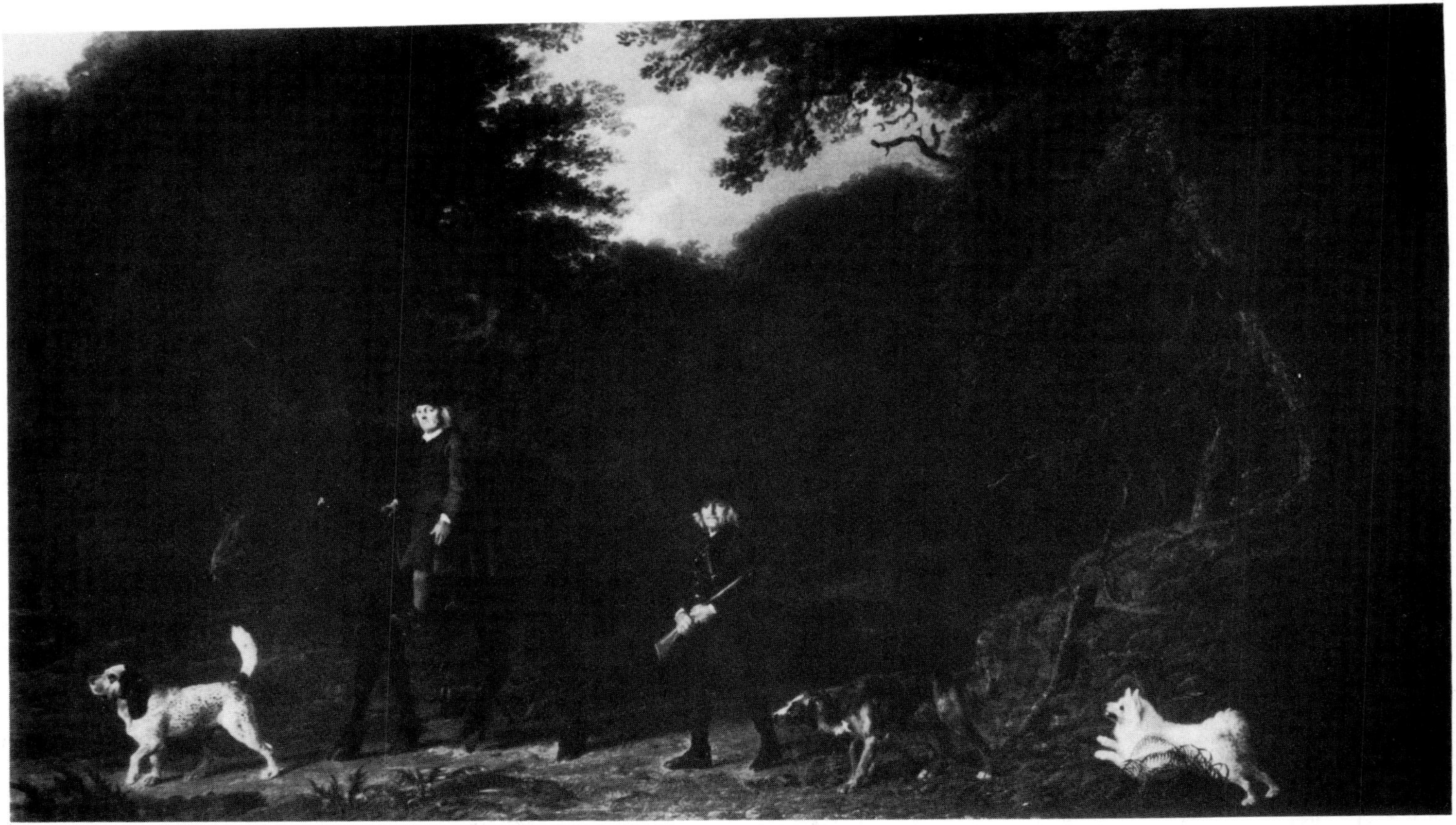

Pl. 10. George Stubbs, *Gamekeepers*
Coll. Lord Iveagh

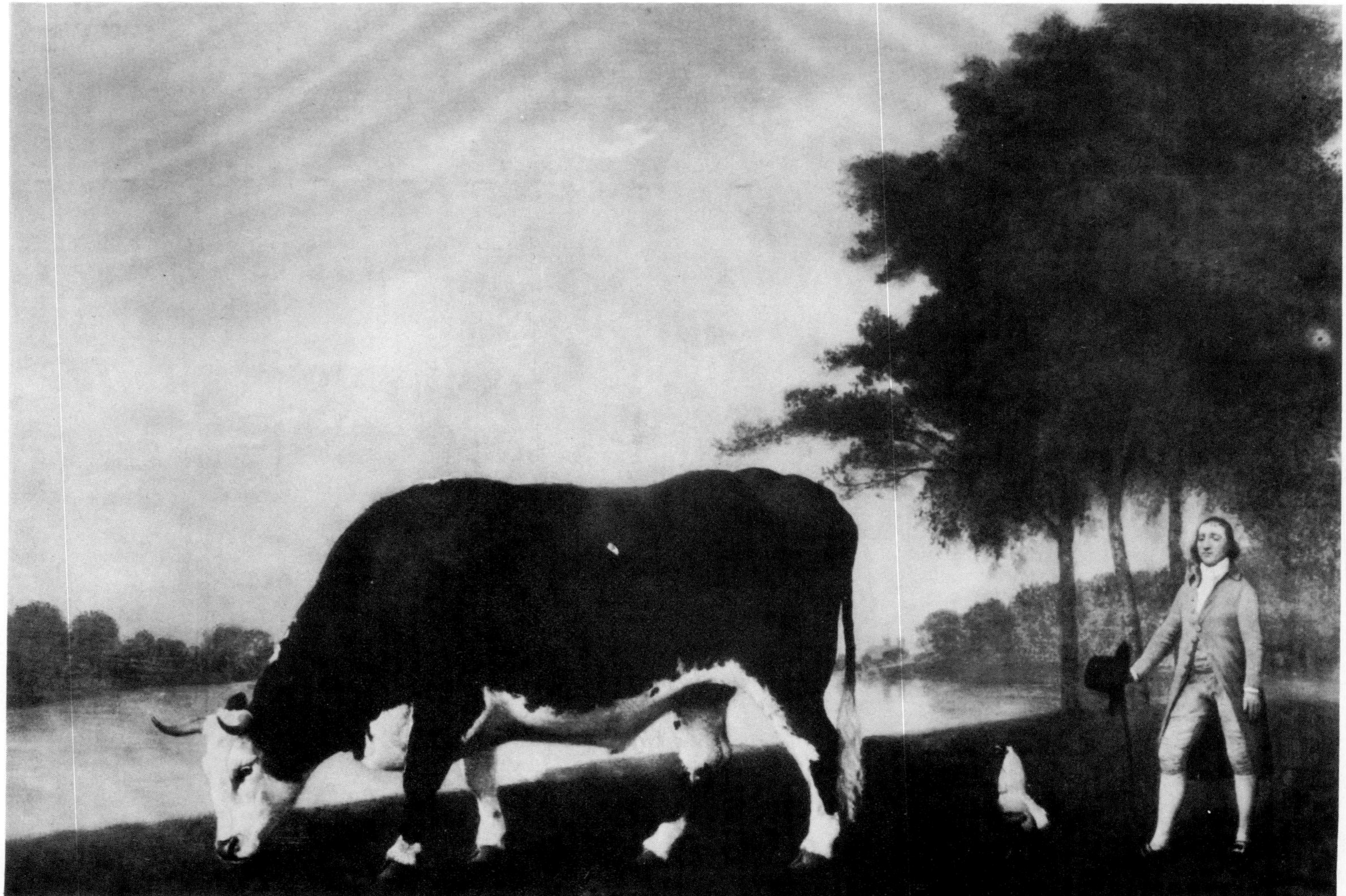

Pl. 11. George Stubbs, *Lincolnshire Ox*
Walker Art Gallery, Liverpool

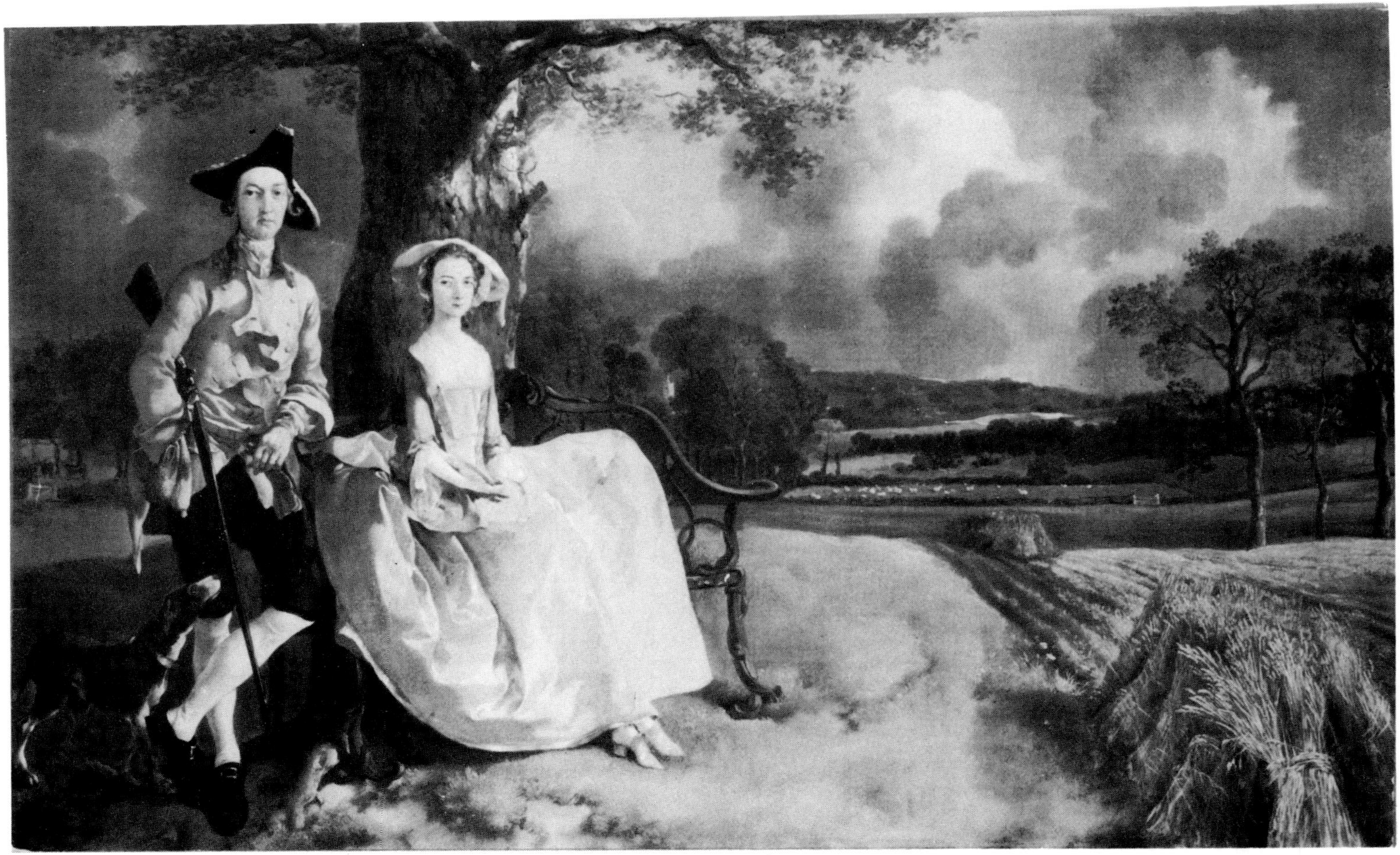

Pl. 12. Thomas Gainsborough, *Portrait of Mr. and Mrs. Andrews*
London, National Gallery

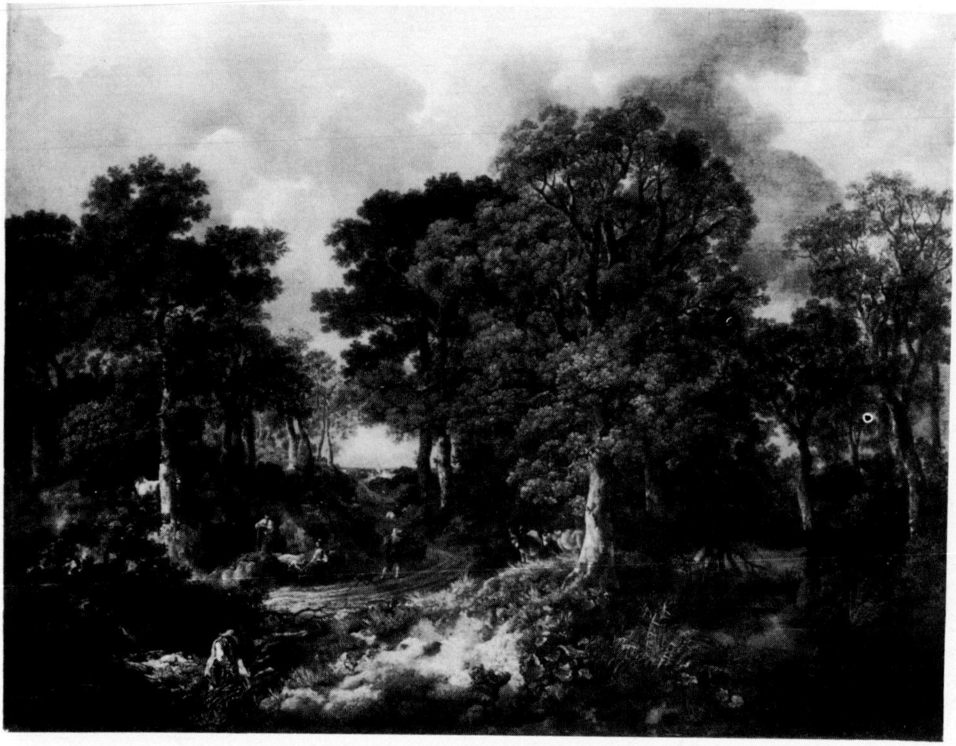

Pl. 13. Thomas Gainsborough, *Cornard Wood*, London, National Gallery

Pl. 14. Thomas Gainsborough, *The Market Cart*, London, Tate Gallery

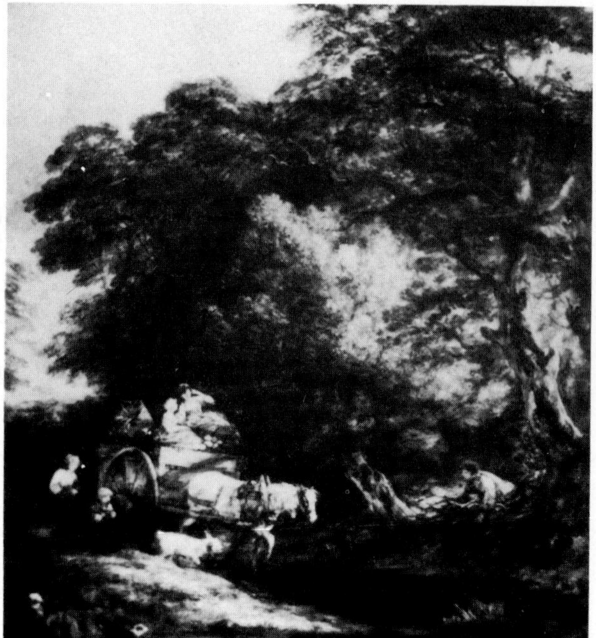

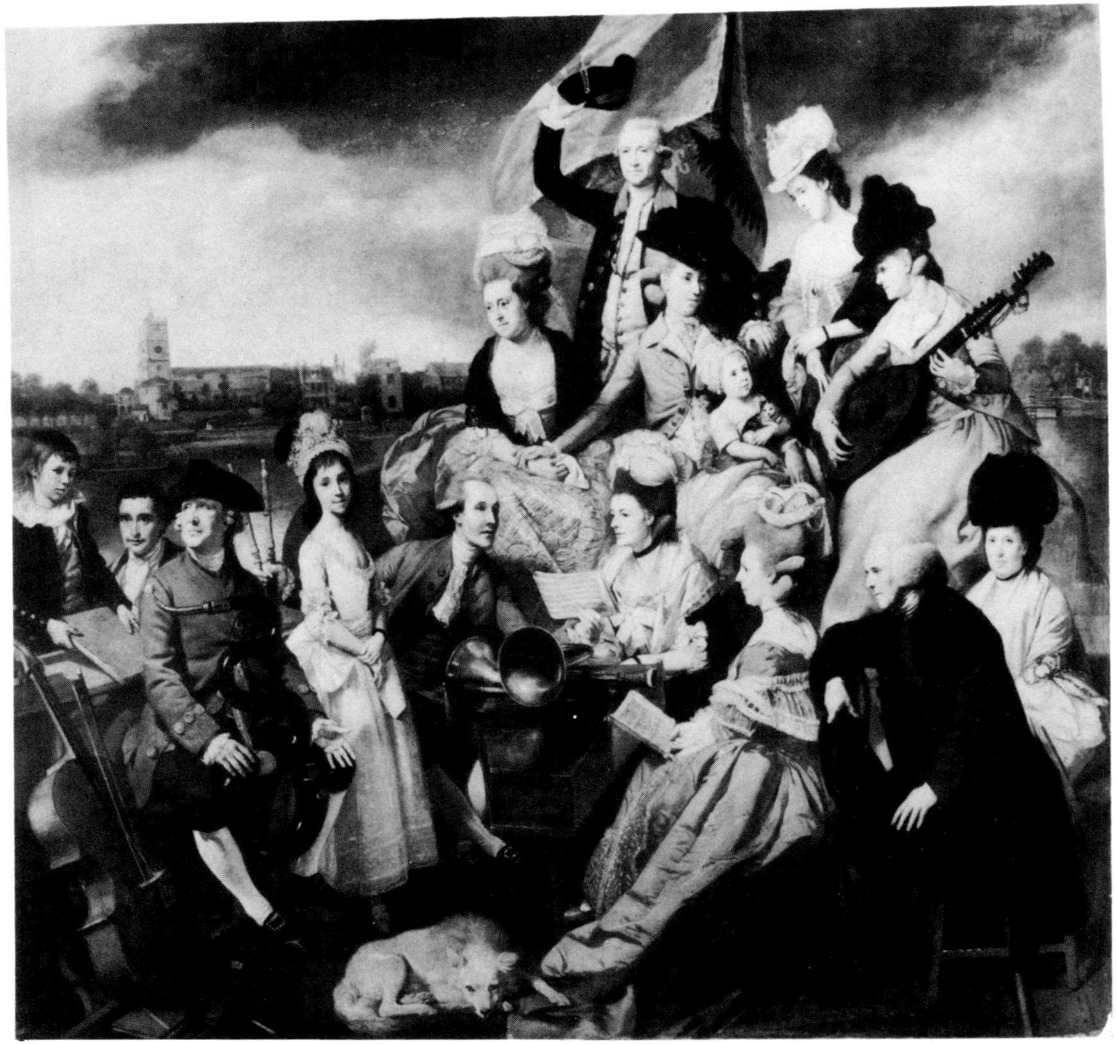

Pl. 15. Zoffany, *The Sharp Family at Fulham*

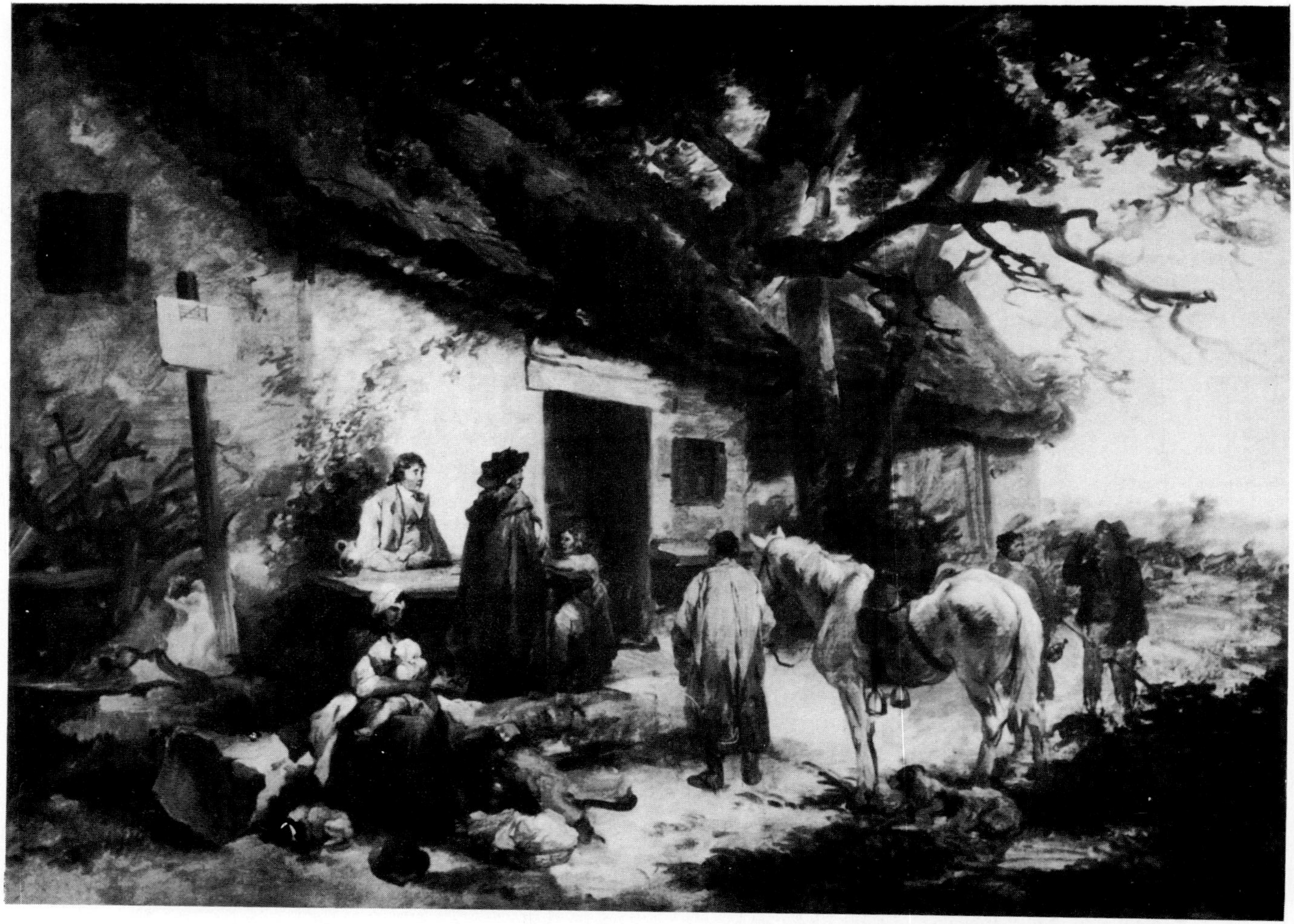

Pl. 16. George Morland, *A Roadside Alehouse*
Lever Collection, Port Sunlight, Liverpool

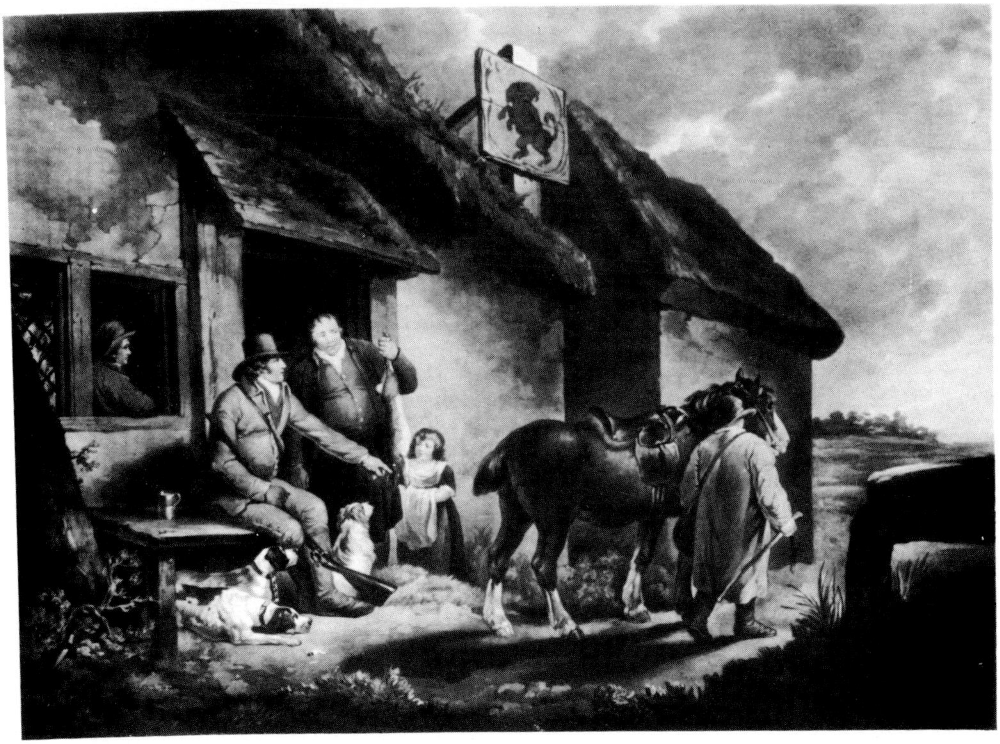

Pl. 17. George Morland, *Evening*

Pl. 18. David Loggan, *Prospect of Cambridge*
From *Cantabrigia Illustrata*, 1650

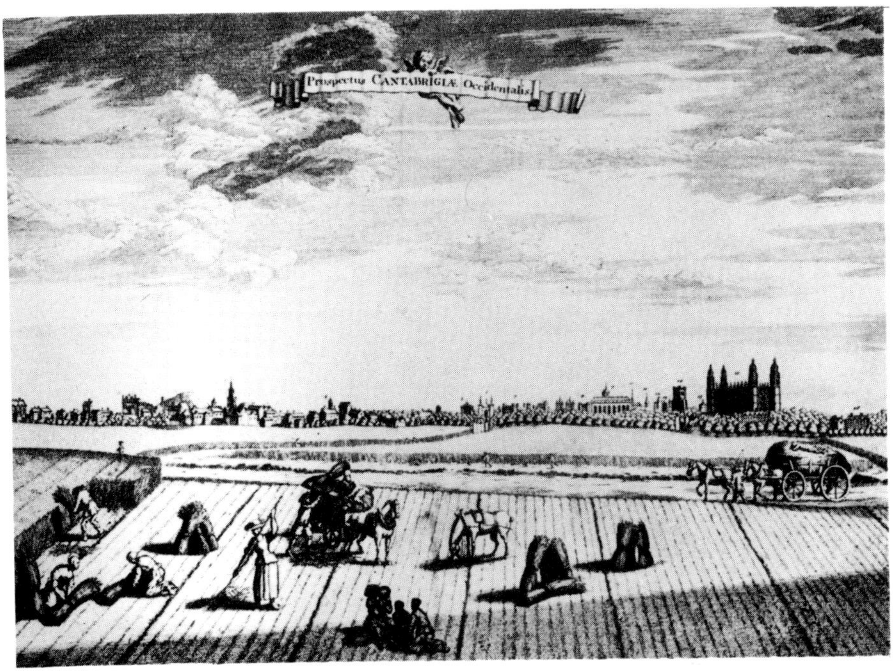

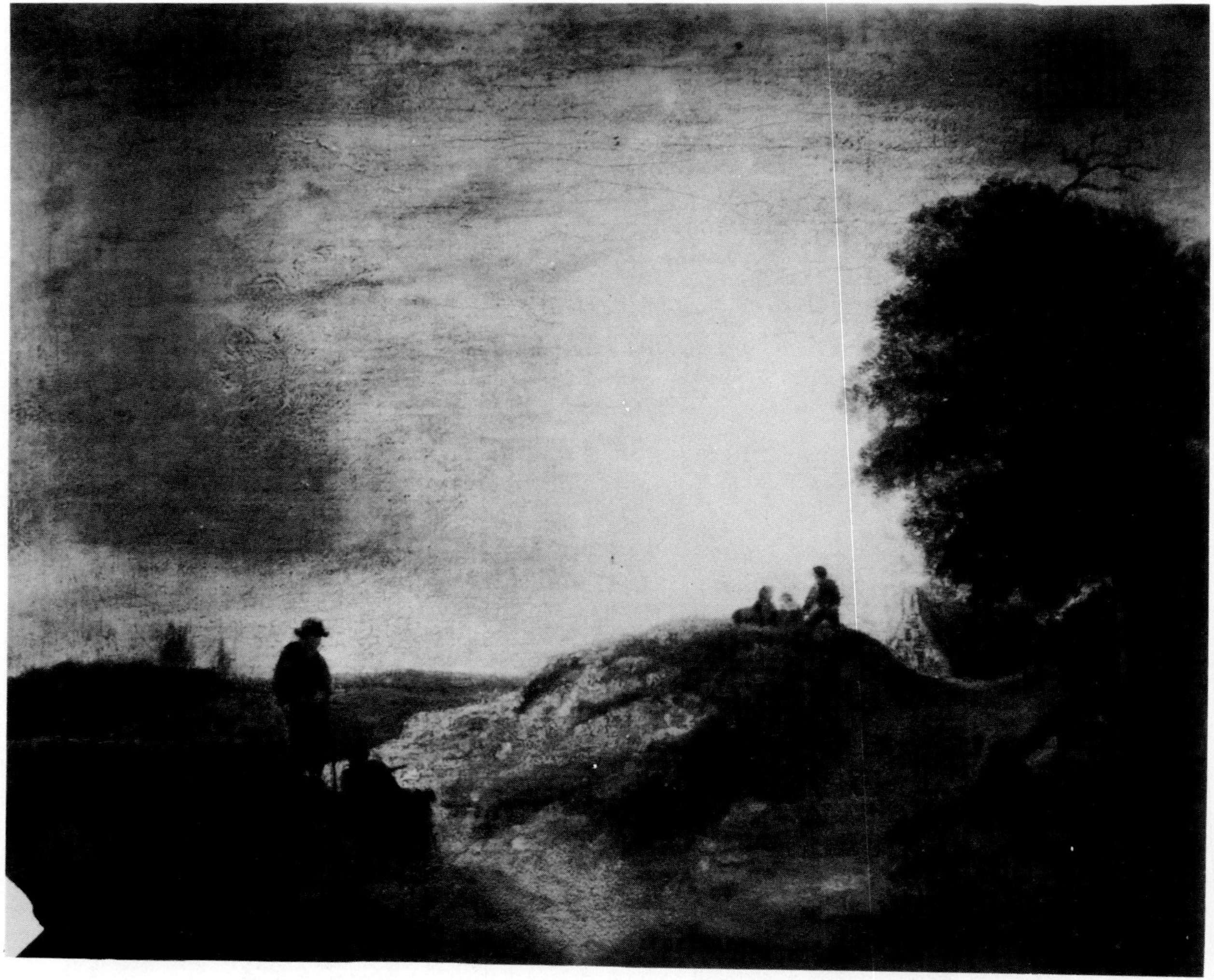

Pl. 19. Norwich School, *Mousehold Heath*
Norwich Castle Museum

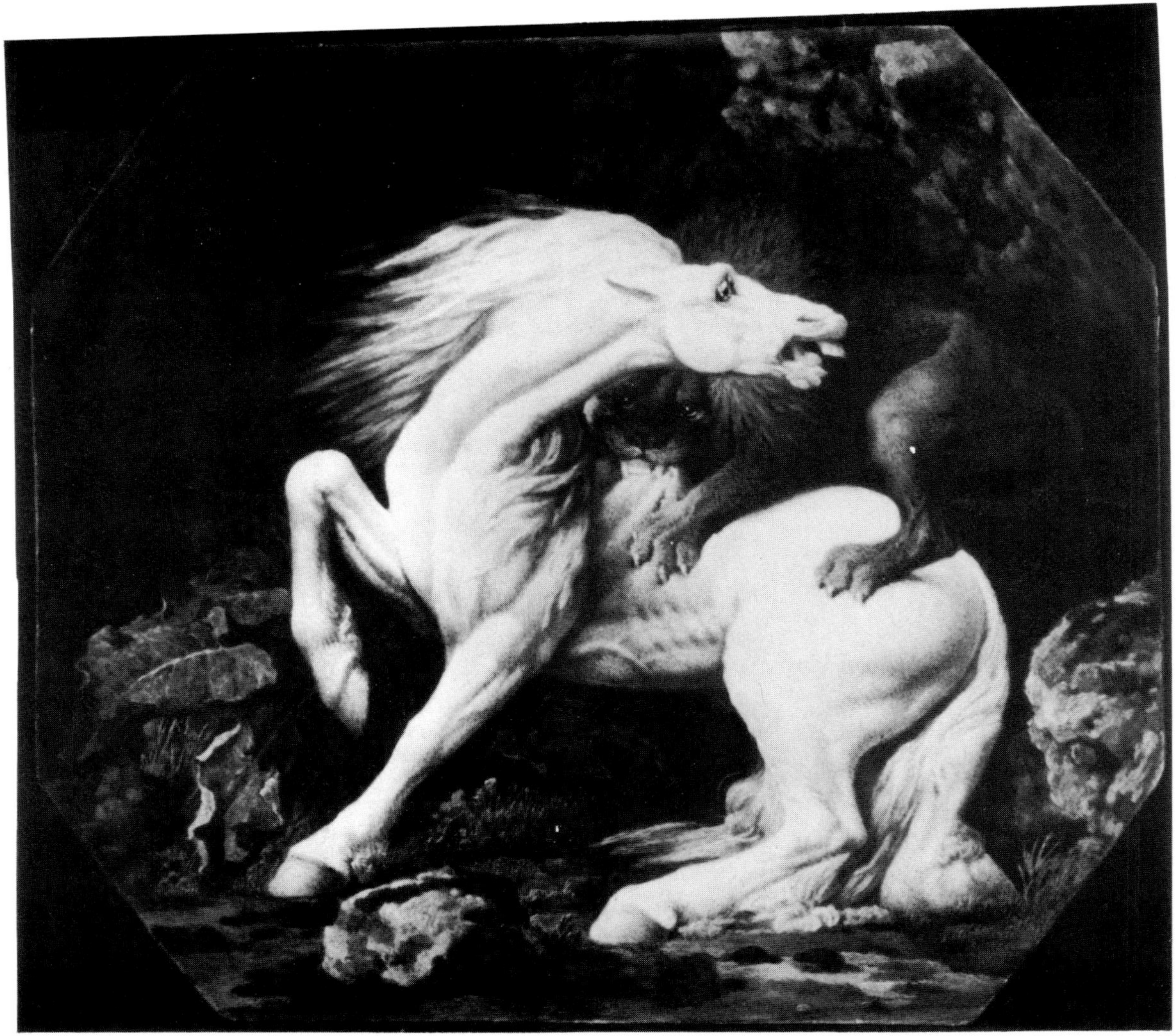

Pl. 20. George Stubbs, *Horse attacked by Lion*
Tate Gallery

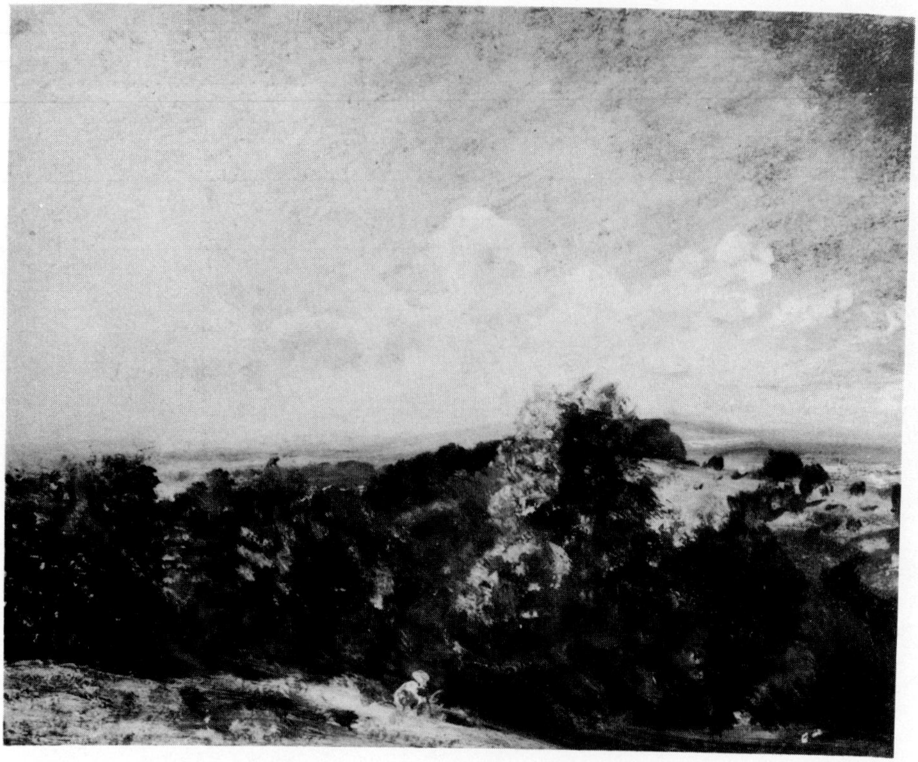

Pl. 21. Constable, *Hampstead Heath*
Royal Academy

Pl. 22. J. W. M. Turner, *Rain, Steam and Speed*
London, National Gallery

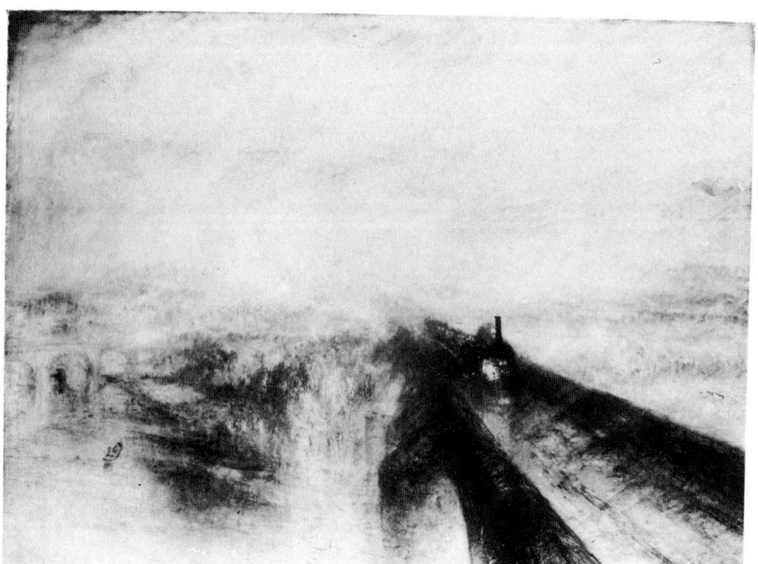